QP 624 NEI

D1766600

DNA STRUCTURE
AND RECOGNITION

This item is in demand. Please
THERE ARE HEAVY FINES

Due for return

IN FOCUS

Titles published in the series:

*Antigen-presenting Cells
*Complement
Cytokines
DNA Replication
DNA Structure and Recognition
DNA Topology
Enzyme Kinetics
Gene Structure and Transcription 2nd edn
Genetic Engineering
Growth Factors
*Immune Recognition
Intracellular Protein Degradation
*B Lymphocytes
*Lymphokines
Membrane Structure and Function
Molecular Basis of Inherited Disease 2nd edn
Molecular Genetic Ecology
Protein Biosynthesis
Protein Engineering
Protein Structure
Protein Targeting and Secretion
Regulation of Enzyme Activity
*The Thymus

*Published in association with the British Society for Immunology.

Series editors

David Rickwood

Department of Biology, University of Essex, Wivenhoe Park,
Colchester, Essex CO4 3SQ, UK

David Male

Institute of Psychiatry, De Crespigny Park, Denmark Hill,
London SE5 8AF, UK

DNA STRUCTURE AND RECOGNITION

Stephen Neidle

Cancer Research Campaign Biomolecular Structure Unit,
The Institute of Cancer Research, Sutton, Surrey SM2 5NG, UK

IRL PRESS
—at—
OXFORD UNIVERSITY PRESS
Oxford New York Tokyo

Oxford University Press, Walton Street, Oxford OX2 6DP

Oxford New York Toronto
Delhi Bombay Calcutta Madras Karachi
Kuala Lumpur Singapore Hong Kong Tokyo
Nairobi Dar es Salaam Cape Town
Melbourne Auckland Madrid
and associated companies in
Berlin Ibadan

Oxford is a trade mark of Oxford University Press

In Focus is a registered trade mark of the Chancellor, Masters, and Scholars
of the University of Oxford trading as Oxford University Press

Published in the United States
by Oxford University Press Inc., New York

© Oxford University Press, 1994

All rights reserved. No part of this publication may be
reproduced, stored in a retrieval system, or transmitted, in any
form or by any means, without the prior permission in writing of Oxford
University Press. Within the UK, exceptions are allowed in respect of any
fair dealing for the purpose of research or private study, or criticism or
review, as permitted under the Copyright, Designs and Patents Act, 1988, or
in the case of reprographic reproduction in accordance with the terms of
licences issued by the Copyright Licensing Agency. Enquiries concerning
reproduction outside those terms and in other countries should be sent to
the Rights Department, Oxford University Press, at the address above.

This book is sold subject to the condition that it shall not,
by way of trade or otherwise, be lent, re-sold, hired out, or otherwise
circulated without the publisher's prior consent in any form of binding
or cover other than that in which it is published and without a similar
condition including this condition being imposed
on the subsequent purchaser.

A catalogue record for this book is available from the British Library

Library of Congress Cataloging in Publication Data
Neidle, Stephen.
DNA structure and recognition / Stephen Neidle.—1st ed.
(In focus)
Includes bibliographical references and index.
1. DNA–Structure. 2. DNA–ligand interactions. I. Title.
II. Series: In focus (Oxford, England)
QP624.N45   1994   574.87'3282–dc20   93–36757
ISBN 0 19 963419 X (pbk.)

QMW LIBRARY
(MILE END)

Typeset by Footnote Graphics, Warminster, Wiltshire
Printed in Malta by Interprint

Preface

The study of DNA structure has been at the forefront of much of structural molecular biology for the past 40 years. It remains an area of intense activity as structural methods continue to reveal levels of complexity and diversity of DNA structures that were unimagined by the discoverers of the double helix. Equally, structural approaches have illuminated the ways in which DNA is recognized by proteins and small molecules. None the less, many major questions remain unresolved, especially the nature of the relationships between DNA sequence and structure at a local level and how drugs and proteins are able to recognize and exploit such features. These issues are at the heart of DNA function itself, and thus of DNA recognition in general.

This book presents an account of the fundamentals of DNA structure (Chapters 2–4) and the principles of DNA recognition (Chapter 5), at a level suitable for undergraduates in the biological, pharmacological, biophysical, and chemical sciences, and for beginning postgraduates in these areas. It is also designed to provide a foundation for subsequent more detailed studies in the fields of DNA–protein and DNA–drug interactions. I have deliberately emphasized results from X-ray crystallographic methods since they continue to make by far the major contribution to our understanding of the structures of DNA/DNA–drug and DNA–protein complexes. It is important that those without crystallographic expertise can none the less be familiar with its basic concepts, and assess the quality and limitations of a particular structure determination. Accordingly, I have included a short section in Chapter 1 on various methods of determining DNA structure, emphasizing X-ray analysis, but summarizing other major biophysical approaches as well.

The DNA structure and recognition field is now a major and many-faceted one. It is inevitable then that in a book of this length, a number of topics are covered only very briefly, if at all; I do hope that the interested reader will be encouraged to pursue such topics further via the lists of reading material and references to the primary and secondary literature.

I am grateful to many colleagues and friends for their help and advice, and for reading sections of the book, especially Helen Berman, Andrew Lane, Charlie Laughton, Christine Nunn, and Mark Sanderson and to Helen Berman, Horace Drew, Laurence Hurley, Tom Steitz, Andrew Lane, Dick

Dickerson, and Dinshaw Patel for generously supplying material and pre-prints. Finally, I thank my wife Andrea and children Daniel, Ben, and Hannah for their patience and encouragement during the writing of this book, and the Cancer Research Campaign for their support of studies on DNA and DNA–drug structures in my laboratory over many years.

Stephen Neidle

Contents

4. DNA–DNA recognition: non-standard DNA structures   57

5. Principles of ligand–DNA recognition   71

To my parents

1

Methods for studying DNA structure

1. Introduction

Our knowledge of DNA structure has come a long way since the elucidation of the double helix in 1953 by Watson and Crick in conjunction with the X-ray fibre diffraction data of Franklin and Wilkins. Many details of the relationships between DNA primary sequence and molecular structure are now understood, in large part from structural studies on defined-sequence oligonucleotides. There is also a considerable body of experimental information on the structures of protein–DNA and drug–DNA complexes. DNA structure itself continues to surprise with its ability to exist in a wide variety of forms, such as left-handed and triple-stranded helices.

The discovery of the double helix, as Watson and Crick realized, immediately provided fundamental new insights into the nature of genetic events. Our more recent knowledge of both the detail and the variety of DNA structures themselves, together with the manner in which they are recognized by regulatory proteins, mutational compounds, and drugs, is starting to pave the way to altogether more profound levels of understanding of the processes of gene regulation, mutation/carcinogenesis, and drug action at the molecular level. It may well be that this knowledge will be exploitable in the future, for example in terms of the artificial and highly specific suppression of particular disease-associated genes.

However, much remains to be discovered. The underlying relationship between DNA sequence, structure, and flexibility are as yet only partially understood, yet are of great significance for the interaction of both small molecules and proteins with DNA and for the function of the resulting complexes. Protein–DNA recognition, even within a particular family of DNA-binding proteins, is much more complex than was thought when the first regulatory protein structures were determined. The ultimate target of the unravelling and generalization of protein–DNA recognition codes still remains elusive.

These advances in DNA structural studies have been largely due to the increased power and sophistication of the experimental approaches of X-ray

1

crystallography, which have provided most of the highly detailed information on DNA structures. The dominance of the crystallographic approach still continues, and is reflected in the balance of this book. Nuclear magnetic resonance (NMR) spectroscopy, molecular modelling/simulation, and chemical/biochemical probe techniques are playing increasingly important roles in providing complementary information on DNA dynamics and flexibility that can approach atomic resolution in at least some of its detail. It should not be forgotten that there have been other major, technical factors in the rapid advance of knowledge on DNA structure and on DNA–protein interactions; these have been the development of routine oligonucleotide synthesis methods, and the advent of efficient cloning and expression systems for DNA-binding proteins. This chapter provides a brief introduction to these biophysical methods, emphasizing their scope and limitations for DNA structural studies.

2. X-ray diffraction methods for structural analysis

2.1 Overview

X-rays typically have a wavelength of the same dimensions as inter-atomic bonds in molecules (about 1.5 Å). Scattering (or diffraction) of X-rays by molecules in ordered matter is the result of interactions between the radiation and the electron distribution of each component atom. Typical diffraction patterns from DNA, in the form of fibres or single crystals, are shown in *Figure 1.1*. Reconstruction of the internal molecular arrangement by analysis of the scattered X-rays, analogous to a lens focusing scattered light from a microscope sample, thus provides a picture of the electron density distribution in the molecule. This reconstruction is not generally straightforward, due to the loss of phase information from the individual reflected X-rays during the diffraction process. The phase problem needs to be solved (see below for a brief description of various methods of doing this) in order for the electron density to be calculated in three dimensions (as a Fourier series), which is commonly termed a Fourier map. The equivalence of X-ray wavelength and bond distance means that in principle, the electron density of individual atoms in a molecule can be resolved, provided that the pattern of diffracted X-rays can be reconstituted into a real-space image. The degree of electron density detail that can actually be seen is dependent on the resolution of the diffraction pattern observed. Resolution may be thus defined in terms of the shortest separation between objects (i.e. atoms or groups of atoms in a molecule) that can be observed in the electron density reconstituted from the diffraction pattern. The resolution limit (r) is governed by the maximum diffraction angle (θ) recorded for the diffraction date and the wavelength (λ) of the X-rays: r is defined as $\lambda/2\sin\theta$. At a resolution of 2.5 Å, individual atoms in a structure cannot be resolved in an electron-density map although the shape and orientation of ring systems can be readily distinguished. They appear as elongated regions of electron density, with substituents being apparent as 'outgrowths' from the main density. At 1.5 Å, individual atoms are generally just observable in a map; and only at about 1.0–1.2 Å are all atoms fully resolved and separated from each other.

The number of individual diffraction maxima observed from a crystalline or semi-crystalline sample depends directly on the resolution of the pattern, with for example, a high-resolution (1.3 Å) structure of an oligonucleotide crystal giving 5000–6000 individual maxima (reflections). The intensity of an individual reflection is proportional to its structure amplitude, or observed structure factor, which when combined with phase information results in the calculated structure factor. X-ray structures are always optimized (refined) against these observed data, usually by a least-squares method (see below). The accuracy and reliability of the resulting structure depends in part on the amount and resolution of the diffraction data, as well as the quality of their measurement. Of key importance is the actual correctness of the structural model itself, both in gross outline (which is defined by the low- to medium-resolution data), and in the detailed aspects of the structure (defined by the high-resolution data). The standard method for assessing these factors is to calculate the crystallographic R factor, defined as: $R = \Sigma\|F_o\| - |F_c\|/\Sigma|F_o|$, where F_o and F_c are the observed and calculated structure factors. Values of R for a correct structure can range from 0.15 to over 0.20; in general the lower the value the more reliable is the model. It is common practice to calculate Fourier maps with parts of a structure omitted to verify that these parts reappear in the map at the correct positions.

2.2 Fibre diffraction methods

Historically DNA structures were first analysed by fibre diffraction methods. They continue to play a minor, though still significant role in DNA structure analysis. Polymeric nucleic acids directly extracted from cell nuclei have not been crystallized as single crystals capable of three-dimensional structure analysis. Instead, they can be readily made into fibres, when the act of 'pulling' such a fibre orients the nucleic acid helix along the direction of the fibre. These fibres can have exactly repetitious helical dimensions even though the underlying DNA sequences in them do not, since the sequence information in them is lost. An exception is with synthetic polynucleotides of known, simply repeating sequence such as poly(dA–dT)·poly(dA–dT). Natural and synthetic polynucleotides can form fibres with varying degrees of internal order, having one- or two-dimensional para-crystalline arrays in the fibre, with the latter usually having the greater order because of their non-random sequences. These differing degrees of order are reflected in their X-ray diffraction patterns, with natural DNA molecules usually having a degree of order along the helix axis but being randomly oriented with respect to each other. This gives rise to an X-ray diffraction pattern with characteristic spots and streaks of intensity—for example, the 'helical cross' diffraction pattern, which is characteristic of B-type double helices. Such patterns can be analysed to give the helical dimensions of pitch, rise, and number of residues per helical turn, as well as defining the overall helical type (A, B, etc.). Even the best-ordered of paracrystalline polynucleotide fibres gives at most only a few hundred individual diffraction maxima, corresponding to a maximum resolu-

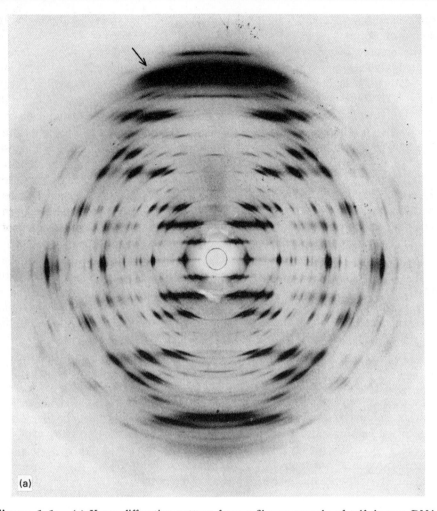

Figure 1.1. (a) X-ray diffraction pattern from a fibrous sample of calf thymus DNA, showing a characteristic B-form pattern. The very strong meridian intensity (shown arrowed) corresponds to an inter-planar spacing of 3.4 Å. (b) X-ray diffraction picture from a single crystal of an oligonucleotide–drug complex (by Dr R. McKenna, Institute of Cancer Research), taken at the SERC Synchrotron Radiation Source using the Laue method.

tion of about 2.5 Å. It is not in general possible to analyse this pattern of diffraction intensities, determine phases, and derive a molecular structure *ab initio*, since the pattern is an average from all of the nucleotide units in a helical repeat. Instead, the pattern is fitted to a model using a least-squares procedure which enables conformational details of the (mono-or di-nucleotide) repeat to be varied and optimized. The correctness and quality of the model may be assessed with the standard crystallographic R factor. Values of 0.15–0.25 indicate that the calculated diffraction pattern agrees well with the observed one, and that the

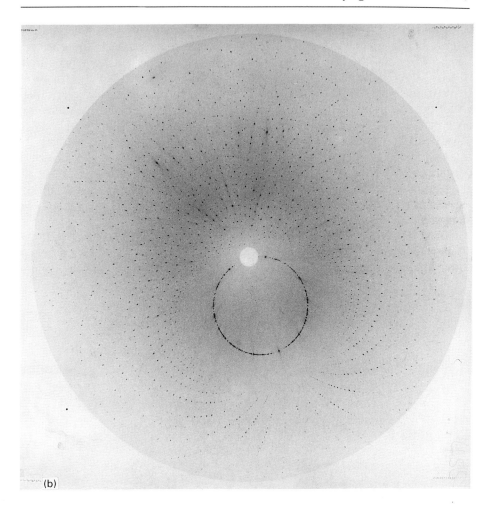

(b)

model is physically reasonable in terms of its stereochemistry. The fact that the process assumes a starting model and that others might in principle fit the data at least moderately well, does not necessarily mean that the phase problem for these fibre structures has been uniquely solved. This led some years ago to several suggestions of alternative structures to the Watson–Crick anti-parallel double helix for DNA. None of these alternatives can be fitted in an acceptable manner to the observed diffraction data, as defined by the R factor and other tests. This, together with their numerous close intramolecular contacts, enabled these alternative structures to be conclusively rejected and the anti-parallel double helix accepted as the sole model that can acceptably fit the observed diffraction data.

2.3 Single-crystal methods

By contrast, single-crystal X-ray crystallographic analyses are able to determine the complete three-dimensional molecular structures of biological macromolecules provided they are discrete and not the effectively infinite polymers of nucleic acid fibres. Single crystals are ordered arrays of discrete molecules in three dimensions. The resolution for single-crystal studies of oligonucleotides ranges from 0.8 to 3.0 Å (*Figure 1.2*). Thus, the highest-resolution oligonucleotide structures have true atomic resolution and accordingly are of corresponding accuracy (≤ 0.1 Å for distances and $\leq 1°$ for angles) in respect of derived geometric parameters. A typical 2.5 Å resolution structure analysis would have distances reliable to about ± 0.3 Å and angles to about $\pm 5°$, although the use of constraints to standard bond geometries means that it is chiefly non-bonded and intermolecular distances that have to be interpreted with care. Hydrogen atoms cannot be located in even the highest-resolution oligonucleotide analyses, and so hydrogen bonding schemes (especially those involving water molecules) can only be inferred.

X-ray diffraction patterns from oligonucleotide crystals can be analysed, and their underlying molecular structures solved *ab initio*, by the standard heavy-atom multiple isomorphous replacement phasing methods of protein crystal-

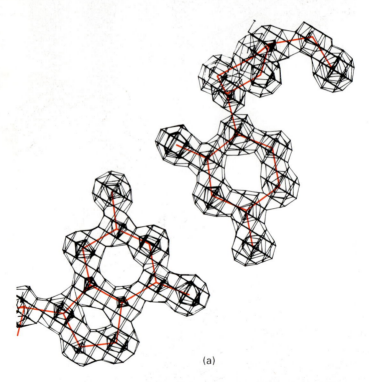

(a)

Figure 1.2.   The electron-density calculated in the plane of a C·G base pair, at various resolutions (a) 1 Å, (b) 1.5 Å, (c) 2 Å, (d) 2.5 Å, and (e) 3 Å. At high resolution, individual atoms are clearly visible, whereas at low resolution, only the overall shape of a base can be seen. Figures courtesy of Professor Helen Berman and Dr Bohdan Schneider (Rutgers University).

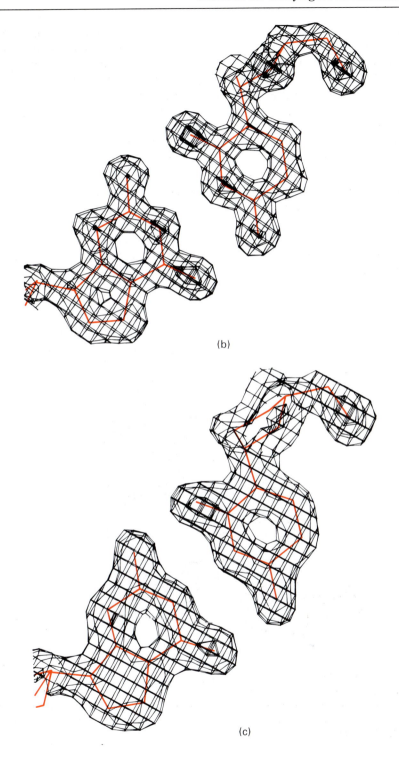

(b)

(c)

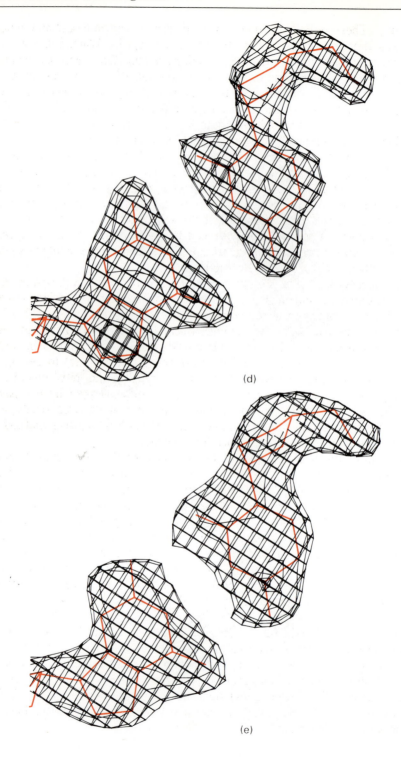

(d)

(e)

lography. These do not presume any particular structural model and hence do not bias the resulting structure to, for example, the Watson–Crick model. Alternatively they can be analysed by taking account of the fact that many such structures are isomorphous to ones solved by heavy-atom phasing or are presumed to contain a particular structural motif such as a double helix. These are normally solved by molecular replacement or 'search' methods, which assume at least part of the structure and attempt to locate it in the crystallographic unit cell. Problems have occasionally arisen with this approach, when, for example, a helix has been correctly oriented within the unit cell but its position is incorrectly located, being systematically related to the correct one by a simple base pair translation. Protein–DNA crystal structures are usually solved by heavy-atom multiple isomorphous replacement methods, in which case these possible ambiguities do not arise. It is fortunate that the key crystal structures of a B- and a Z-DNA oligonucleotide have been solved by these methods (see Chapter 3), thereby ensuring a firm basis for subsequent molecular replacement analyses of a large number of oligonucleotide structures.

Macromolecular crystal structures are normally optimized with respect to the diffraction data by non-linear least-squares fitting procedures, which minimize the differences between observed and calculated structure factors. This is the process of crystallographic refinement, which also uses information from established stereochemical and structural features (such as bond lengths and angles, planar geometry of the DNA bases, preferred torsion angles) to set up constraints and restraints between them and so improve the initial models. The technique of simulated annealing has been recently adopted from molecular dynamics as an effective way of refining structures when large scale (>1 Å) atomic movements are required, since conventional least-squares methods are inherently incapable of effecting such large changes.

Oligonucleotide and oligonucleotide–protein crystals are heavily hydrated, with often over 50 per cent solvent. It is typical for only a small fraction of these water molecules to be located in electron density maps, largely because their high mobility smears their electron density to below the signal-to-noise level of these maps. The majority of water molecules reported in these structures are, unsurprisingly, the least mobile ones, which are directly hydrogen bonded to the structure—these are the 'first-shell' water molecules. The ways in which molecules pack in the crystal are sometimes of importance when examining structural features, since considerations of efficient packing can force parts of molecules to interact one with another and consequently possibly modify some features.

The quality and reliability of an oligonucleotide crystal structure are not straightforward to assess, especially for a non-crystallographer. Yet, judgements on these factors are critical when undertaking and using structural comparisons and analyses. The important crystallographic parameters of quality have been outlined above. Of at least equal significance are the derived stereochemical features—examination of these is also a reliable guide to quality. Particular features to examine include:

(a) close non-bonded intra- and intermolecular contacts that are less than the sum of the van der Waals radii of the atoms involved;

(b) the distribution of values for torsion angles around single bonds (eclipsed (~ 0 °) values are indicative of problems in refinement);

(c) hydrogen bonds with distances appreciably outside the accepted ranges of ~ 2.7–3.2 Å.

3. NMR methods for studying DNA structure and dynamics

Nuclear magnetic resonance (NMR) methods enable structures to be determined in solution, largely by means of measurements of proton–proton coupling constants and through-space nuclear Overhauser effect (NOE) derived distances. This has the obvious advantage that molecules do not have to be crystallized, which is often the major (and highly frustrating!) limitation to the study of a macromolecule by X-ray crystallography. There is also the apparent advantage that a structure determined in solution is more relevant to physiological processes than an X-ray crystallographic study in the solid state. However, the two techniques should not be considered as alternatives. Rather they are complementary, providing distinct information. For example, NMR results emphasize the flexible nature of DNA molecules and the fact that individual groups such as sugars are dynamically in motion. It is gratifying that parallel observations of sequence-dependent effects in a number of oligonucleotide sequences have been reported from both crystallographic and NMR studies, although differences in detail are sometimes apparent.

There are a number of limitations to the accuracy and reliability of NMR methods as applied to nucleic acids. The NOE is only significant for protons, and so phosphate geometry is not directly defined by it. Since NOE intensity is proportional to the inverse sixth power of a proton–proton distance, it is a short-range effect (much less than 5 Å) and longer distances important in DNA structure, such as groove width, may not be derived directly. Most reliability from NMR experiments can be placed on qualitative features of DNA structure, especially base pairing, sugar pucker, and location of ligand-binding sites. Detailed aspects of, in particular sequence-dependent structure, are still a matter of considerable controversy. This is in large part because of the relative lack of NOE data as compared with the several thousands of X-ray intensities from a typical crystallographic analysis. The consequent under-determination of NMR-derived structures means that their effective 'resolution' is probably ~ 4 Å, with parts of a structure providing the NOE data being the most reliable. This situation is distinct from that for small (<25 kDa) globular proteins, where the richness of NOE data arising from the compact, closely packed amino acid residues, provides for highly reliable and detailed NMR structure determination more equivalent to that from high-resolution crystallographic analyses.

NMR coupling constants between protons are related to the dihedral angle between them. Hence their measurement provides direct information on sugar puckers and part of the backbone conformation in a nucleotide or oligonucleotide.

4. Molecular modelling and simulation of DNA

Crystallographic analyses provide a quasi-static view of DNA structure. The process of X-ray data acquisition from a single crystal takes upwards of 24 h. It thus provides a time-averaged picture of molecular motions about the low-energy structure in the crystal (which is typically more than 50 per cent solvent). By contrast, molecular modelling techniques enable dynamic changes in structure and conformation to be calculated and visualized in terms of their effects on molecular energetics. The theoretical methods thus provide information complementary to the experimental techniques.

It is not feasible to compute conformational or energetic properties for DNA sequences quantum-mechanically. Instead, force-fields have been derived from experimental data that describe the energetics of a DNA molecule in terms of the sum of a number of factors: van der Waals non-bonded interactions, bond length and angle distortions, barriers to rotation about bonds, electrostatic contributions from full and partial electrostatic-potential derived atomic charges, and hydrogen bonding. Nucleic acid force-fields do not as yet adequately reflect the contributions of polarization effects, which can be important in some circumstances. The major empirical nucleic acid force-fields have been incorporated into algorithms that minimize the conformation of a molecule with respect to its internal energy. This is the method of molecular mechanics, which in effect optimizes local low-energy minima. Much more extensive explorations of conformations can be made by molecular dynamics, which applies Newton's equations of motion to an empirical force-field, for all atoms in a molecule. This technique is very computer-intensive and realistic simulations of molecular motions, especially when large numbers of calculated solvent molecules are included, are currently restricted to at most a few hundred picoseconds of molecular movements. Molecular dynamics can overcome barriers between local minima, unlike molecular mechanics, and is being widely used in conjunction with distance geometry data (from NOEs) to derive plausible DNA and DNA–protein structures from these measurements. The use of molecular mechanics and dynamics methods has greatly increased in recent years, largely due to their widespread availability as packaged computer programs, often with graphical front-ends for the display of results. There are three such programs in common use that have their force-fields parameterized for nucleic acids and their components:

(a) AMBER ('Assisted Model Building and Energy Refinement'), from the laboratory of P. A. Kollman, University of California, San Francisco;

(b) CHARMM ('Chemistry at Harvard Macromolecular Mechanics'), from the laboratory of M. Karplus, Harvard University;

(c) GROMOS ('Groningen Molecular Simulation'), from the laboratory of W. F.

van Gunsteren and H. J. C. Berendsen, University of Groningen, The Netherlands.

These are fully described in the references given at the end of this chapter.

5. Chemical and enzymatic probes of structure

Enzymes such as DNase I cleave the phosphodiester bond in a DNA duplex at every nucleotide position, although the cutting efficiency is markedly dependent on sequence, and by implication, on sequence-related structural features. Cleavage is blocked by protein or drug binding. Hence DNase I can be used to determine sites of binding along a DNA sequence as well as to assess possible effects of particular sequences on DNA structure. Chemical cleaving agents such as hydroxyl radicals can give similar information. Since these are much smaller molecules than cleavage enzymes, their effects on DNA structure are less perturbing and sequence-dependent. Other types of chemical probe can attack specific base sites; these can be useful in defining the precise sites of protection resulting from drug or protein binding to a DNA sequence.

These methods have the important advantage over the fine-structure techniques of crystallography and NMR, namely that they are applicable to long (up to several thousand base pair) DNA sequences, and are thus more directly relevant to DNA in the cell. Hence, the use of chemical and enzymatic probes for DNA provides a way of obtaining at least some data at the molecular level on otherwise inaccessible structural problems in DNA–protein and DNA–drug recognition.

6. Sources of structural data

The results of a crystal structure or fibre diffraction analysis are most useful as a set of atomic coordinates. Those from fibre diffraction are available either in the primary literature or in various review chapters and compilations (see Chapter 3). Crystallographic coordinates may be obtained from a database, provided that they have been deposited in the first instance! The majority of available oligonucleotide crystal structures are in the Brookhaven Protein Data Bank; many are accompanied by structure factor data. The Cambridge UK Crystallographic Database (which is primarily for small molecules), also contains coordinate data on a number of oligonucleotide structures. A more recent development is the establishment, at Rutgers University, USA, of a comprehensive relational database solely for nucleic acid crystallographic data—the Nucleic Acid Database. This has an almost complete compilation of coordinates, and provides for the accession of both primary and derived data by means of computer networks. It provides a set of powerful tools for the comparative study of nucleic acid structural features.

7. Further reading

General:

Bates, A. D. and Maxwell, A. (1993). *DNA topology.* Oxford University Press. Covers higher-order DNA structure–complementary to this volume.

Blackburn, G. M. and Gait, M. J. (eds) (1990). *Nucleic acids in chemistry and biology.* IRL Press, Oxford. A useful survey of nucleic acid function, with some emphasis on structural aspects.

Calladine, C. R. and Drew, H. R. (1992). *Understanding DNA.* Academic Press, London. An entertaining account of selected areas of DNA structure.

Eckstein, F. and Lilley, D. M. J. (eds) (1987 onwards). *Nucleic acids and molecular biology,* Vols 1–6. Springer-Verlag, Berlin. Contains reviews of many important areas, emphasizing molecular biological relevance.

Lilley, D. M. J. and Dahlberg, J. E. (eds) (1992) *DNA structures.* Vols 211 and 212 of *Methods in Enzymology.* Academic Press, San Diego. A comprehensive compilation of reviews on many aspects of DNA structure, emphasizing methodologies.

Olby, R. (1974). *The path to the double helix.* Macmillan Press, London. An authoritative and detailed account of the history of the discovery of the double helical structure.

Saenger, W. (1984). *Principles of nucleic acid structure.* Springer-Verlag, Berlin. An essential text, providing a wealth of background information and detail.

Watson, J. D. (1968). *The double helix.* Weidenfeld and Nicolson, London. A highly personalized and entertaining account—essential reading for all budding students of DNA structure.

X-ray crystallography:

Arnott, S. (1970). *Progress in Biophysics and Molecular Biology,* **21**, 265.

Dickerson, R. E. (1992). *Methods in Enzymology,* **211**, 67.

Glusker, J. P. and Trueblood, K. N. (1985). *Crystal structure analysis: a primer,* (2nd edn). Oxford University Press, London.

Hahn, M. and Heinemann, U. (1993). *Acta Crystallographica,* **D49**, 468.

NMR methods:

Lane, A. N. (1989). *Molecular and Cellular Biology,* **8**, 53.

Patel, D. J., Shapiro, L., and Hare, D. R. (1987). *Quarterly Review of Biophysics,* **20**, 1.

van de Ven, J. M. and Hilbers, C. W. (1988). *European Journal of Biochemistry,* **178**, 1.

Wemmer, D. E. (1991). *Current Opinion in Structural Biology,* **1**, 452.

Wüthrich, K. (1986). *NMR of proteins and nucleic acids.* Wiley, New York.

Molecular modelling and simulation methods:

Berendsen, H. J. C. (1991). *Current Opinion in Structural Biology,* **1**, 191.

Dean, P. M. (1987). *Molecular foundations of drug–receptor interactions.* Cambridge University Press, Cambridge, UK.

Goodfellow, J. M. and Williams, M. A. (1992). *Current Opinion in Structural Biology,* **2**, 211.

McCammon, J. A. and Harvey, S. C. (1987). *Dynamics of proteins and nucleic acids.* Cambridge University Press, Cambridge, UK.

van Gunsteren, W. F. and Berendsen, H. J. C. (1990). *Angewandte Chemie,* **29**, 992.

Chemical and enzymatic probes of DNA structure:

Tullius,T.D. (1989). In *Nucleic acids and molecular biology*, (ed. F.Eckstein and D.M.J.Lilley), Vol. 3, pp. 1–12. Springer-Verlag, Berlin.
Tullius, T.D. (1991). *Current Opinion in Structural Biology*, **1**, 428.

Nucleic acid databases:

Berman,H.M., Olson,W.K., Beveridge,D.L., Westbrook,J., Gelbin,A., Demeny,T., Hsieh,S.-H., Srinivasan,A.R., and Schneider,B. (1992). *Biophysical Journal*, **63**, 751.

2

The building-blocks of DNA

1. Introduction

Chemical degradation studies in the early years of this century on material extracted from cell nuclei established that the high molecular weight 'nucleic acid' was actually composed of individual acid units, termed nucleotides. Four distinct types were isolated—guanylic, adenylic, cytidylic, and thymidylic acids. These could be further cleaved to phosphate groups and four distinct nucleosides. The latter were subsequently identified as consisting of a deoxypentose sugar and one of four nitrogen-containing heterocyclic bases. Thus, each repeating unit in a nucleic acid polymer consists of these three units linked together—a phosphate group, a sugar, and one of the four bases. These latter are planar aromatic heterocyclic molecules and are divided into two groups—the pyrimidine bases thymine and cytosine and the purine bases adenine and guanine. Their major tautomeric forms are shown in *Figure 2.1*. Thymine is replaced by uracil in ribonucleic acids, which also have an extra hydroxyl group at the 2' position of their (ribose) sugar groups. The standard nomenclature for the atoms in nucleic acids, as approved by the International Union of Biochemistry, is shown in *Figures 2.1* and *2.2*. Accurate bond length and angle geometries for all bases and nucleosides have been well established by X-ray crystallographic analyses. These values have been incorporated in the force-fields used in molecular mechanics and dynamics modelling, in order to define equilibrium values for the geometric parameters. Accurate crystallographic analyses, at very high resolution, can also directly yield quantitative information on the distribution of electron density in a molecule, and hence on individual partial atomic charges. These charges for nucleosides have hitherto been obtained by *ab initio* quantum mechanical calculations, but are now available experimentally for all four DNA nucleosides (1).

Individual nucleoside units are joined together in a nucleic acid in a linear manner, through phosphate groups attached to the 3' and 5' positions of the sugars (*Figure 2.2*). Hence the full repeating unit in a nucleic acid is a 3',5'-nucleo**tide**.

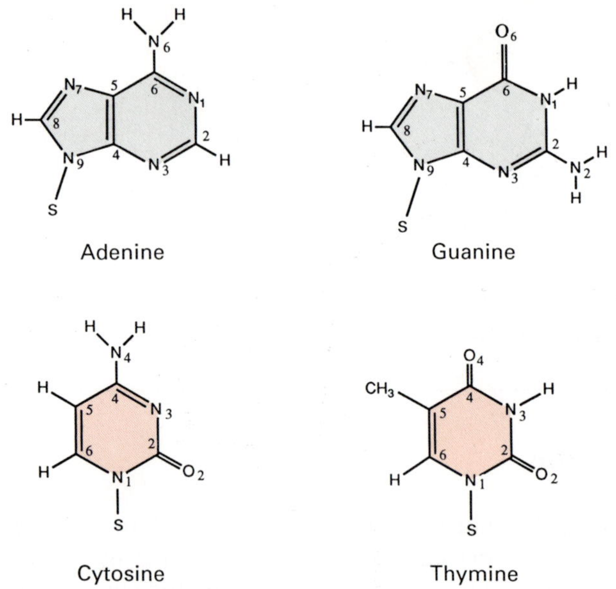

Figure 2.1.   The four bases of DNA.

In nucleic acid and oligonucleotide sequences it is standard practice to use single-letter abbreviations for the four unit nucleotides—A, T, G, and C. The two classes of bases can be abbreviated as Y or py (pyrimidine) and R or pu (purine). Phosphate groups are usually designated as p. A single oligonucleotide chain is conventionally numbered from the 5′ end, for example, ApGpCpTpTpG has the 5′ terminal adenosine nucleoside, with a free hydroxyl at its 5′ position and thus the 3′ end guanosine has a free 3′ terminal hydroxyl group. Intervening phosphate groups are sometimes omitted when a sequence is written down. Chain direction is sometimes emphasized with 5′ and 3′ labels. Thus an anti-parallel double-helical sequence can be written as:

5′-CpGpCpGpApApTpTpCpGpCpG
3′-GpCpGpCpTpTpApApGpCpGpC

or simply (because this example is a palindromic sequence) as

(CGCGAATTCGCG)$_2$.

Publications concerning DNA structure usually prefix a sequence with 'd' (eg dCGAT) to emphasize that the oligonucleotide contains deoxyribose sugars rather than ribose ones.

The bond between sugar and base is known as the glycosidic bond. Its stereochemistry is important. In natural nucleic acids the glycosidic bond is always β, that is the base is above the plane of the sugar when viewed on to the plane and therefore on the same face of the plane as the 5′ hydroxyl substituent

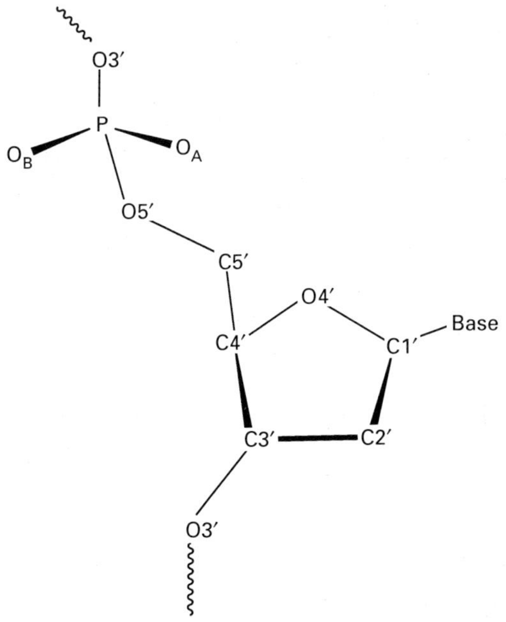

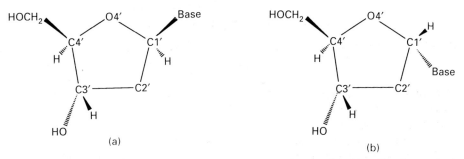

Figure 2.2.   The organization of repeating units in a polynucleotide chain.

(*Figure 2.3a*). The absolute stereochemistry of other substituent groups on the deoxyribose sugar ring of DNA is defined such that, when viewed end-on with the sugar ring oxygen atom O4′ at the rear (*Figure 2.3a*), the hydroxyl group at the 3′ position is below the ring and the hydroxymethyl group at the 4′ position is above. A unit nucleotide can have its phosphate group attached at either the 3′ or the 5′ end, and is thus termed either a 3′ or a 5′ nucleotide. It is chemically possible to construct α-nucleosides and from them α-oligonucleosides, which have their bases in the 'below' configuration relative to the sugar rings and their other substituents (*Figure 2.3b*). These are much more resistant to nuclease

Figure 2.3.   (a) The stereochemistry of a natural β-nucleoside. Bonds marked (▬) are coming out from the plane of the page, towards the reader. Bonds marked (.........) are going away from the reader. (b) Stereochemistry of an α-nucleoside.

attack than standard natural β-oligomers and have been used as antisense oligomers to mRNAs on account of their superior intracellular stability.

2. Base pairing

The realization that the planar bases can associate in particular ways by means of hydrogen bonding was a major step in the elucidation of the structure of DNA. The important early experimental data of Chargaff showed that the molar ratios of adenine:thymine and cytosine:guanine in DNA were both unity. This led to the proposal by Crick and Watson that in each of these pairs the purine and pyrimidine bases are held together by specific hydrogen bonds, to form planar base pairs. In native, double-helical DNA the two bases in a base pair necessarily arise from two separate strands of DNA (with intermolecular hydrogen bonds) and so hold the DNA double helix together (2).

The adenine:thymine (A·T) base pair has two hydrogen bonds while the guanine:cytosine (G·C) base pair has three (*Figure 2.4*). Fundamental to the Watson–Crick arrangement is that the sugar groups are both attached to the bases on the same side of the base pair. As will be seen in Chapter 3, this defines the mutual positions of the two sugar–phosphate strands in DNA itself. The two base pairs are required to be almost identical in dimensions by the Watson–Crick model. High resolution (0.8–0.9 Å) X-ray crystallographic analyses of the ribo-dinucleoside monophosphate duplexes $(GpC)_2$ and $(ApU)_2$ by A. Rich and colleagues in the early 1970s (3, 4) has established accurate geometries for these A·T and G·C base pairs (*Table 2.1*). These structure determinations showed that there are only small differences in size between the two types of base pairs, as indicated by the distance between glycosidic carbon atoms in a base pair. The C1′ . . . C1′ distance is 10.67 Å in the G·C base pair structure, and 10.48 Å in the A·U-containing dinucleoside.

3. Sugar puckers

The five-membered deoxyribose sugar ring in DNA is inherently non-planar. This non-planarity is termed puckering. The precise conformation of a deoxyribose ring can be completely specified by the five endocyclic torsion angles within it (*Figure 2.5*). The ring puckering arises from the effect of non-bonded interactions between substituents at the four ring carbon atoms—the energetically most stable conformation for the ring has all substituents as far apart as possible. Thus different substituent atoms would be expected to produce differing types of puckering. The puckering can be described by either a simple qualitative description of the conformation in terms of atoms deviating from ring coplanarity, or precise descriptions in terms of the ring internal torsion angles.

In principle, there is a continuum of interconvertible puckers, separated by energy barriers. These various puckers are produced by systematic changes in

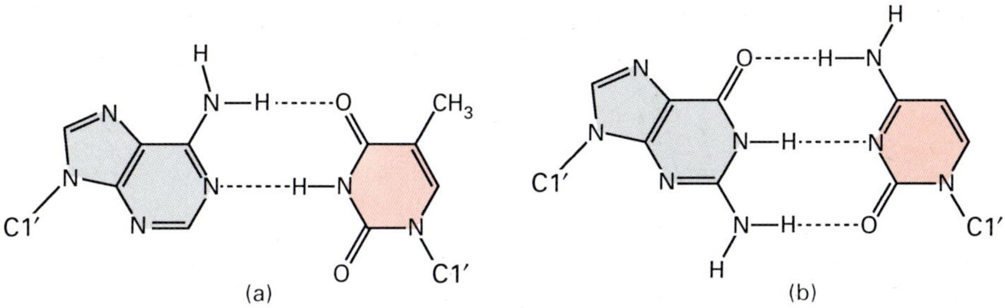

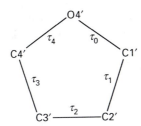

Figure 2.4.   (a) A·T and (b) G·C base pairs, showing Watson–Crick hydrogen bonding.

Table 2.1   Hydrogen-bond distances in Watson–Crick base pairs (2, 3)

	Hydrogen-bonded atoms	Distance (Å)[a]
U·A:	N3-H. N1	2.835 (8)
	O4. . . . .H-N6	2.940 (8)
C·G:	O2. . . . .H-N2	2.86 (1)
	N3. . . . .H-N1	2.95 (1)
	N4-H. O6	2.91 (1)

[a] Estimated standard deviations are in parentheses.

Figure 2.5.   The five internal torsion angles in a deoxyribose ring.

the ring torsion angles. The puckers can be succinctly defined by the parameters P and τ_m (5). The value of P, the phase angle of pseudorotation, indicates the type of pucker since P is defined in terms of the five torsion angles τ_0–τ_4:

$$\tan P = \frac{(\tau_4 + \tau_1) - (\tau_3 + \tau_0)}{2 \times \tau_2 \times (\sin 36° + \sin 72°)}$$

and the maximum degree of pucker, τ_m, by

$$\tau_m = \frac{\tau_2}{\cos P}$$

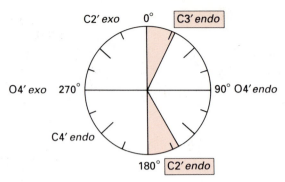

Figure 2.6.   The pseudorotation wheel for a deoxyribose sugar. The shaded areas indicate the preferred ranges of the pseudorotation angle for the two principal sugar conformations.

The pseudorotation phase angle can take any value between 0° and 360°. If τ_2 has a negative value, then 180° is added to the value of P. The pseudorotation phase angle is commonly represented by the pseudorotation wheel, which indicates the continuum of ring puckers (*Figure 2.6*). Values of τ_m indicate the degree of puckering of the ring: typical experimental values from crystallographic studies on mononucleosides are in the range 25–45°. The five internal torsion angles are not independent of each other, and so to a good approximation any one angle, τ_j, can be represented in terms of just two variables:

$$\tau_j = \tau_m \cos [P + 0.8\pi(j - 1)]$$

A large number of distinct deoxyribose ring pucker geometries have been observed experimentally, by X-ray crystallography and NMR techniques. When one ring atom is out of the plane of the other four, the pucker type is an 'envelope' one. More commonly, two atoms deviate from the plane of the other three, with these two either side of the plane. It is usual for one of the two atoms to have a larger deviation from the plane than the other in this, the 'twist' conformation. The direction of atomic displacement from the plane is important. If the major deviation is on the same side as the base and C4'–C5' bond, then the atom involved is termed *endo*. If it is on the opposite side, it is called *exo*. The most commonly observed puckers in crystal structures of isolated nucleosides and nucleotides are close to either C2'*endo* or C3'*endo* types (*Figure 2.7a* and *b*). In practice, these pure envelope forms are rarely observed, largely because of the differing substituents on the ring. Consequently the puckers are then best described in terms of twist conformations. When the major out-of-plane deviation is on the *endo* side, there is a minor deviation on the opposite, *exo* side. The convention used for describing a twist deoxyribose conformation is that the major out-of-plane deviation is followed by the minor one, for example C2'*endo*, C3'*exo*. The C2'*endo* family of puckers have P values in the range 140–185°; in view of their position on the pseudorotation wheel, they are sometimes termed S (south) conformations. The C3'*endo* domain has P values in the range −10 to +40°, and can be termed N (north).

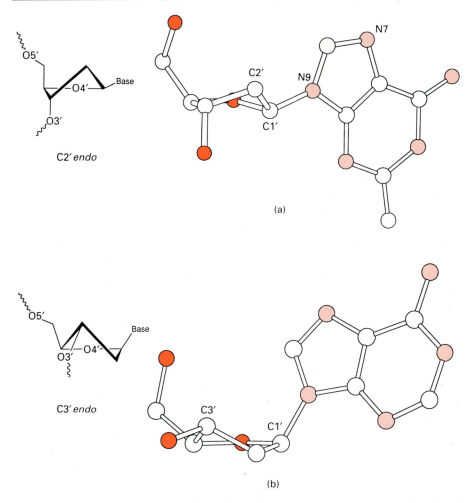

(a)

(b)

Figure 2.7. (a) C2′ *endo* sugar puckering for the guanosine nucleoside, viewed along the plane of the sugar ring. (b) C3′ *endo* sugar puckering for the adenosine nucleoside, with the sugaring viewed in the same direction as in panel (a).

The pseudorotation wheel implies that deoxyribose puckers are free to inter-convert. In practice, there are energy barriers between major forms. The exact size of these barriers has been the subject of considerable study (6, 7). The consensus is that the barrier height is dependent on the route around the pseudorotation wheel. For interconversion of C2′ *endo* to C3′ *endo* the preferred pathway is via the O4′ *endo* state, with a barrier of 8–21 kJ/mol found from an analysis of a large body of experimental data (6), and a somewhat smaller (potential energy) value of 6.3 kJ/mol from a molecular dynamics study (8). The former value, being an experimental one, represents the total free energy for interconversion.

Relative populations of puckers can be monitored directly by NMR measurements of the ratio of coupling constants between H1'–H2' and H3'–H4' protons. These show that in contrast to the 'frozen out' puckers found in the solid state structures of nucleosides and nucleotides, there is rapid interconversion in solution. Nonetheless, the relative populations of the major puckers are dependent on the type of base attached. Purines show a preference for the C2'*endo* pucker conformational type whereas pyrimidines favour C3'*endo*. Deoxyribose nucleosides are primarily (>60%) in the C2'*endo* form, while ribonucleosides favour C3'*endo*. The latter are significantly more restricted in their mobility; this has significance for the structures of oligoribonucleotides. These differences in puckering equilibria and hence in their relative populations in solution and in molecular dynamics simulations, are reflected in the patterns of puckers found in surveys of crystal structures (9). Again, this is a demonstration of the complementarity of information provided by the different structural techniques. Sugar pucker preferences have their origin in the non-bonded interactions between substituents on the sugar ring, and to some extent on their electronic characteristics. For example, the C3'*endo* pucker (*Figure 2.7b*) would have hydroxyl substituents at the 2' and 3' positions further apart than with C2'*endo* pucker; hence the preference of the former by ribonucleosides.

Correlations have been found, from crystallographic and NMR studies, between pucker and several backbone conformational variables, both in isolated nucleosides/nucleotides and in oligonucleotide structures. These are discussed later in this chapter. Changes in sugar pucker are important in oligo- and polynucleotides because they can alter the orientation of C1', C3', and C4' substitutents, resulting in major changes in backbone conformation and overall structure, as indeed is found (Chapter 3). Sugar pucker is thus an important determinant of oligo- and polynucleotide conformation.

4. Conformations about the glycosidic bond

The glycosidic bond links a deoxyribose sugar and a base, being the C1'–N9 bond for purines and the C1'–N1 bond for pyrimidines. The torsion angle χ around this single bond can in principle adopt a wide range of values, although as will be seen, structural constraints result in marked preferences being observed. Glycosidic torsion angles are defined in terms of the four atoms O4'–C1'–N9–C4 (for purines) or O4'–C1'–N1–C2 (for pyrimidines). Theory has predicted two principal low-energy domains for the glycosidic angle, in accord with experimental findings for a large number of nucleosides and nucleotides. The *anti* conformation has the N1,C2 face of purines and the C2,N3 face of pyrimidines directed away from the sugar ring (*Figure 2.8a*) so that the hydrogen atoms attached to C8 of purines and C6 of pyrimidines are lying over the sugar ring. Thus, the Watson–Crick hydrogen bonding groups of the bases are directed away from the sugar ring. These orientations are reversed in the *syn* conformation, with these hydrogen bonding groups now oriented towards the sugar and

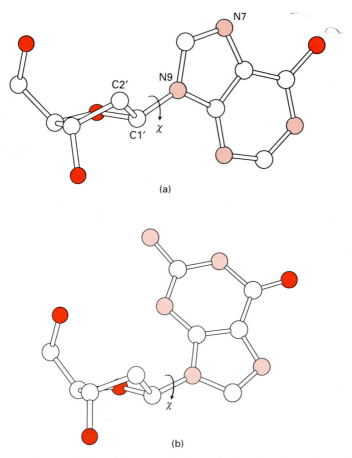

(a)

(b)

Figure 2.8.   (a) The *anti* glycosidic angle conformation for the adenosine nucleoside, with the angle indicated by the curved arrow. (b) The *syn* glycosidic angle conformation for guanosine.

especially its O5′ atom (*Figure 2.8b*). Analyses of a number of crystal structures of *syn* purine nucleosides have found hydrogen bonding between the O5′ atom and the N3 base atom, which would stabilize this conformation. Otherwise, for purines, the *syn* conformation is slightly less preferred than the *anti*, on the basis of fewer non-bonded steric clashes in the latter case. The principal exceptions to this rule are guanosine-containing nucleotides, which have a small preference for the *syn* form because of favourable electrostatic interactions between the exocyclic N2 amino group of guanine and the 5′-phosphate atom. For pyrimidine nucleotides, the *anti* conformation is preferred over the *syn*, because of unfavourable contacts between the O2 oxygen atom of the base and the 5′-phosphate group. The results of molecular-mechanics energy minimizations on all four DNA nucleotides in both *syn* and *anti* forms (using the AMBER all-atom force-field), are fully in accord with these observations (*Table 2.2*).

Table 2.2 Energy differences between glycosidic angle conformers for B-form DNA nucleotides, calculated with the AMBER program (10)

Nucleotides	Energy difference (kJ/mol)
A(*anti*) > A(*syn*)	1.3
T(*anti*) > T(*syn*)	7.1
G(*syn*) > G(*anti*)	13.9
C(*anti*) > C(*syn*)	7.6

The sterically preferred ranges for the two domains of glycosidic angles are: $-120 < \chi < 180°$ for *anti* and $0 < \chi < 90°$ for *syn*. Values of χ in the region of about $-90°$ are often described as 'high *anti*'. There are pronounced correlations between sugar pucker and glycosidic angle, which reflect the changes in non-bonded clashes produced by C2'*endo* versus C3'*endo* puckers. Thus, *syn* glycosidic angles are not found with C3'*endo* puckers due to steric clashes between the base and the H3' atom, which points towards the base in this pucker mode.

5. The backbone torsion angles and correlated flexibility

The phosphodiester backbone of an oligonucleotide has six variable torsion angles (*Figure 2.9*), designated $\alpha \ldots \zeta$, in addition to the five internal sugar torsions $\tau_0 \ldots \tau_4$ and the glycosidic angle χ. As will be seen, a number of these have highly correlated values (and therefore correlated motions in a solution environment). Steric considerations alone dictate that the backbone angles are restricted to discrete ranges (11, 12) (*Figure 2.10*), and are accordingly not free to adopt any value between 0 and 360°. *Figure 2.10* uses a conformational wheel to show these preferred values, which are directly readable from their positions around the wheel. The fact that angles α, β, γ, and ζ each have three allowed ranges, together with the broad range for angle ε that includes two staggered regions, leads to a large number of possible low-energy conformations for the unit nucleotide, especially when glycosidic angle and sugar pucker flexibility are taken into account. In reality, only a few oligonucleotide and polynucleotide structural classes have actually been observed out of this large range of possibilities; this is doubtless in large part due to the restraints imposed by Watson–Crick base pairing on the backbone conformations when two DNA strands are intertwined. In contrast, crystallographic and NMR studies on a large number of standard and modified mononucleosides and nucleotides have shown their considerably greater conformational diversity, in accord with the possibilities indicated in *Figure 2.10*. For these, backbone conformations in the solid state and in solution are not always in agreement; the requirements for efficient packing in the crystal can often overcome the modest energy barriers between different values for a particular torsion angle.

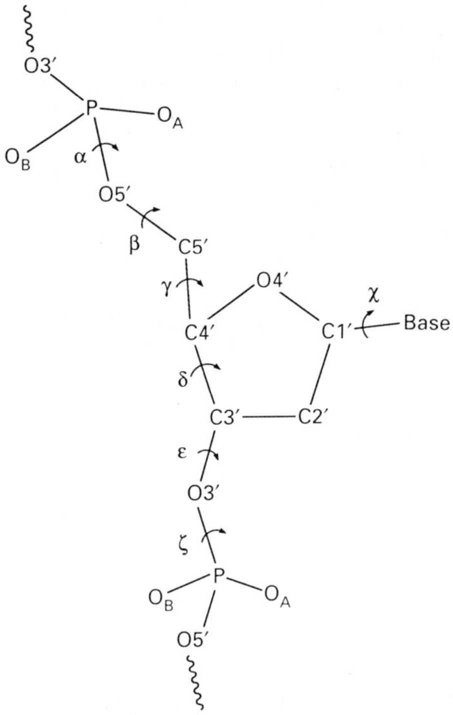

Figure 2.9.   The backbone torsion angles in a unit nucleotide; each rotatable bond is indicated by a curved arrow.

A common convention for describing these backbone angles is to term values of $\sim 60°$ as *gauche*$^+$ (g$^+$), those of about $-60°$ as *gauche*$^-$ (g$^-$), and those of $\sim 180°$ as *trans* (t). Thus, for example, angles α (about the P–O5' bond), and γ (the exocyclic angle about the C4'–C5' bond), can be in the g$^+$, g$^-$, or t conformations. The two torsion angles around the phosphate group itself, α and ζ, have been found to show a high degree of flexibility in various dinucleoside crystal structures, with the tg$^-$, g$^-$g$^-$ and g$^+$g$^+$ conformations all having been observed (13). As will be described in Chapter 3, the g$^-$g$^-$ conformation can place successive nucleotide units in arrangements that have their bases in potential hydrogen bonding positions with respect to a second nucleotide strand. Thus, this is the phosphate conformation for DNA (and RNA) right-handed double helices. The torsion angle β, about the O5'–C5' bond, is almost always found to be *trans*. All three possibilities for the γ angle have been observed in nucleoside crystal structures, although the g$^+$ conformation predominates in right-handed oligo- and polynucleotide double helices. The *trans* conformation for γ places the 5'-phosphate group in quite a distinct position with respect to the deoxyribose ring. The torsion angle δ around the C4'–C3' bond adopts values that relate to the pucker of the sugar ring, since the internal ring torsion angle,

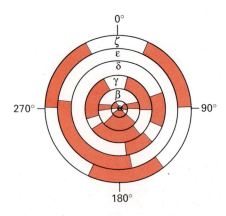

Figure 2.10. The allowed ranges of backbone torsion angles (shown shaded) in nucleosides, nucleotides, oligo- and polynucleotides, shown as a conformational wheel.

τ_3, (also around this bond), has a value of ~35° for C2′*endo* and about −40° for C3′*endo* puckers. δ is ~75° for C3′*endo* and ~150° for C2′*endo* puckers.

An alternative nomenclature for torsion angles ranges used by some in the DNA field is that common in organic chemistry (the Klyne–Prelog system). In this, the *syn* (s) designation is given to angles clustered around 0°, and the *anti* (a) designation to those around 180°. Intermediate angles are defined as ±*synclinal* (±*sc*) for around ±60°, and ± *anticlinical* (±ac) for around ±120°.

There are a number of well-established correlations involving pairs of these backbone torsion angles, as well as sugar pucker and glycosidic angle. These have been observed in mononucleosides and nucleotides (which are inherently more flexible in solution as well as being more subject to packing forces in the crystal), and more recently, in oligonucleotides. Studies of oligonucleotide crystal structures has revealed some further conformational inter-relationships. These are discussed in Chapter 3. The existence of such correlations is important: it means that the flexibility of, and the atomic motions in oligo- and polynucleotides follow concerted patterns of inter-dependence (see Chapter 3). In general, these correlations are due to the diminution of non-bonded contacts that occur with particular conformations, as described in Section 4. Some of the more important correlations that have been observed in mononucleosides and nucleotides are as follows.

1. There is a correlation between sugar pucker and glycosidic angle χ, especially for pyrimidine nucleosides. C3′*endo* pucker is usually associated with median-value *anti* glycosidic angles, whereas C2′*endo* puckers are commonly found with high *anti* χ angles. *Syn* glycosidic angle conformations show a marked preference for C2′*endo* sugar puckers.

2. The C4′–C5′ torsion angle γ is correlated with the glycosidic angle and to some extent with sugar pucker. *Anti* glycosidic angles tend to correlate with g⁺ conformations for γ.

6. Further reading

Jeffrey, G.A. and Saenger, W. (1991). *Hydrogen bonding in biological structures.* Springer-Verlag, Berlin.
Voet, D. and Rich, A. (1970). *Progress in Nucleic Acid Research and Molecular Biology,* **10**, 183.

7. References

1. Pearlman, D.A. and Kim, S.-H. (1990). *Journal of Molecular Biology,* **211**, 171.
2. Watson, J.D. and Crick, F.H.C. (1953). *Nature,* **171**, 737.
3. Seeman, N.C., Rosenberg, J.M., Suddath, F.L., Kim, J.P., and Rich, A. (1976). *Journal of Molecular Biology,* **104**, 109.
4. Rosenberg, J.M., Seeman, N.C., Day, R.O., and Rich, A. (1976). *Journal of Molecular Biology,* **104**, 145.
5. Altona, C. and Sundaralingam, M. (1972). *Journal of the American Chemical Society,* **94**, 8205.
6. Olson, W.K. and Sussman, J.L. (1982). *Journal of the American Chemical Society,* **104**, 270.
7. Olson, W.K. (1982). *Journal of the American Chemical Society,* **104**, 278.
8. Harvey, S.C. and Prabhakaran, M. (1986). *Journal of the American Chemical Society,* **108**, 6128.
9. Murray-Rust, P. and Motherwell, S. (1978). *Acta Crystallographica,* **B34**, 2534.
10. Weiner, S.J., Kollman, P.A., Nguyen, D.T., and Case, D.A. (1986). *Journal of Computational Chemistry,* **7**, 230.
11. Sundaralingam, M. (1969). *Biopolymers,* **7**, 821.
12. Olson, W.K. (1982). In *Topics in nucleic acid structure, Part 2* (ed. S. Neidle), pp. 1–79. Macmillan Press, London.
13. Kim, S.-H., Berman, H.M., Seeman, N.C., and Newton, M.D. (1973). *Acta Crystallographica,* **B29**, 703.

3

DNA structure as observed in fibres and crystals

---

1. Structural fundamentals

1.1 Helical parameters

The fibre diffraction method for determining the structure of polymeric DNA has been outlined in Chapter 1. Examination of X-ray diffraction photographs of oriented fibres enables the basic parameters defining the dimensions of the helix to be obtained. Measurement of the spacing between the layer lines directly gives the reciprocal of the helical pitch P, defined as the distance, parallel to the helix axis, between successive nucleotide units per complete turn of the helix (*Figure 3.1*). The presence of a meridional reflection for a fibre oriented perpendicular to the X-ray direction, indicates the presence of a structural periodicity in that direction. Analysis of the meridional reflection gives the regular repeat distance d (the helical rise if the nucleotide units are perpendicular to the helix axis). The unit repeat n, the number of nucleotide units in a single full turn of helix, that is per pitch P, is thus P/d. Helical rise is more generally the distance between successive nucleotide units, projected to be parallel to the helix axis if the units are not strictly perpendicular to this axis. The diameter of the helix can be derived from the dimensions of the closely packed unit cell projection down the fibre axis.

1.2 Morphological features of base pairs

Base pairs are not necessarily planar, and indeed are rarely so. This is in part because the geometric requirements of hydrogen bonding are not especially stringent, and in particular the angle subtended at the hydrogen atom (donor . . . H acceptor) can deviate by up to about 35° from linearity without appreciable loss of hydrogen bond energy. Other factors, such as the need to avoid steric clash in some base pair . . . base pair non-bonded interactions within a helix, can result in distortions of base pairs from planarity, since these do not distort hydrogen bonds beyond these energetically favourable limits. These deviations from planarity can take place in a number of ways (1) that can be categorized in two groups of local base pair morphological features. The first category involves individual base

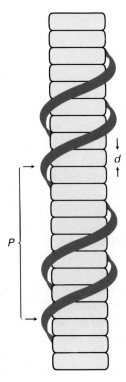

Figure 3.1. A schematic representation of a polynucleotide double helix, with the backbone shown as shaded ribbons and the base pairs as disks. The helical pitch is the distance between the pair of arrows on the left-hand side, and the helical rise is between the small arrows on the right-hand side of the helix.

pairs. It specifies both movements of one base relative to another within a base pair (for example, propeller twist and buckle), and movement of the base pair relative to global axes in the helix. The second category is concerned with the relative movements of base pairs in successive base pairs (base pair steps)—for example, helical twist and roll. It defines the relative relationships of successive base pairs in a structure. Inevitably the second set of features incorporate the first, so correlations between them can be expected. Further detail is given in the Appendix.

2. Polynucleotide structures from fibre diffraction studies

2.1 Classic DNA structures

Information on the dimensions of polynucleotide double helices was initially derived from the diffraction patterns of DNA fibres at a high (92%) relative humidity, which showed a readily interpretable and characteristic Maltese cross pattern of strong intensities. Examination of this pattern, which arises from the

B-DNA double helix, led directly to the original Watson–Crick model in 1953 (2). Subjecting DNA fibres to other conditions produces rather different diffraction patterns. In particular, lower (65–75 per cent) relative humidity results in the A-DNA pattern, which typically has many more diffraction maxima, indicating greater crystallinity and hence order in the fibres. A number of other forms have subsequently been found, most of which are sub-classes of the A and B double helices, the two most important ones for random sequence DNA. DNA is thus highly polymorphic, the differing forms corresponding to distinct yet interconvertible molecular structures. Some DNA structural types have only been found with defined-sequence polynucleotides rather than with random-sequence DNA (the standard A and B forms can occur with most sequences). Under appropriate experimental conditions, the diffraction pattern of such a polynucleotide is best fitted to a repeating unit, which rather than being a simple mononucleotide, can be a dinucleotide or even an oligonucleotide. However, in these instances, the relatively small number of observed diffraction intensities and hence their low ratio compared with the number of structural variables, can make the resulting structures less reliable than those of standard A and B helices.

B-DNA, the classic structure described by Watson and Crick, has subsequently been refined (3) using the linked-atom least-squares procedure developed by Arnott and his group. The helical repeat is a single nucleotide unit, with necessarily all of the nucleotides in the structure having the conformation of this unit. The backbone conformation (detailed in the Appendix) has C2'endo sugar puckers and high anti glycosidic angles (Figure 3.2). The right-handed double helix has 10 bp

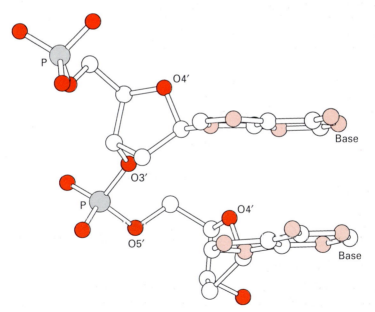

Figure 3.2. Conformation of two successive nucleotides in one strand of the B-DNA conformation, derived from fibre diffraction analysis.

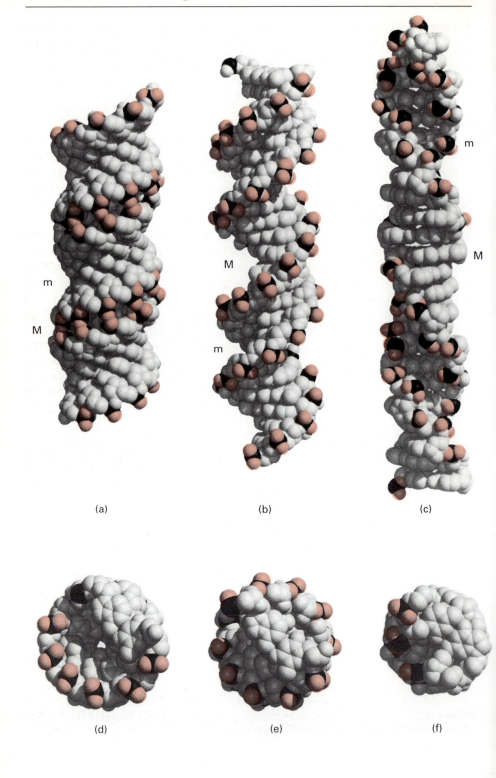

(a)   (b)   (c)

(d)   (e)   (f)

Figure 3.3. (a–c) Space-filling representations of A, B, and Z double helices, with the same number of base pairs in each. This emphasizes the differences in length per complete helical turn. Phosphorus atoms are shaded black. Major (M) and minor (m) grooves are indicated. (d–f) Views of these helices, looking down the helix axes.

per complete turn, with the two polynucleotide chains anti-parallel to each other (*Figure 3.3b*) and linked by Watson–Crick A·T and G·C base pairs. The paired bases are almost exactly perpendicular to the helix axis and they are stacked over the axis itself (*Figure 3.3e*). Consequently the base pair separation is the same as the helical rise, i.e. 3.4 Å. An important consequence of the Watson–Crick base pairing arrangement is that the two deoxyribose sugars linked to an individual base pair are on the same side of it. So, when successive base pairs are stacked on each other in the helix, the gap between these sugars forms continuous indentations in the surface that wind along, parallel to the sugar–phosphodiester chains. These indentations are termed grooves. The asymmetry in the base pairs results in two parallel types of groove, whose dimensions (especially their depths) are related to the distances of base pairs from the axis of the helix and their orientation with respect to the axis. The wide major groove (*Figure 3.3b* and *Table 3.1*) is almost identical in depth to the much narrower minor groove, which has the hydrophobic hydrogen atoms of the sugar groups forming its walls. In general, the major groove is richer in base substituents—O6, N6 of purines and N4, O4 of pyrimidines—than the minor one. This, together with the steric differences between the two, has important consequences for interaction with other molecules. Base pair and base step morphologies for several polynucleotide helices are given in the Appendix.

The A-DNA duplex also has a single nucleotide unit as the helical repeat. This has C3'*endo* sugar puckers, which brings consecutive phosphate groups on the nucleotide chains closer together (5.9 Å, compared with 7.0 Å in B-DNA) and alters the glycosidic angle from high *anti* to *anti*. As a consequence, the base

Table 3.1   Groove dimensions in DNA double helices

Polymorph	Major groove		Minor groove	
	Width (Å)	Depth (Å)	Width (Å)	Depth (Å)
A	2.2	13.0	11.1	2.6
B	11.6	8.5	6.0	8.2
C	10.5	7.6	4.8	7.9
D	9.6	6.2	0.8	7.4
Z	8.8	3.7	2.0	13.8

The data are taken from fibre-diffraction analyses (taken from ref. 3 for the A–D helices and from the singles-crystal data (ref. 44) for the Z helix). Groove width is defined as the perpendicular distance between phosphate groups on opposite strands, minus the van der Waals diameter of a phosphate group (5.8 Å). Groove depths are defined in terms of the differences in cylindrical polar radii between phosphorus and N2 guanine or N6 adenine atoms, for minor and major grooves respectively.

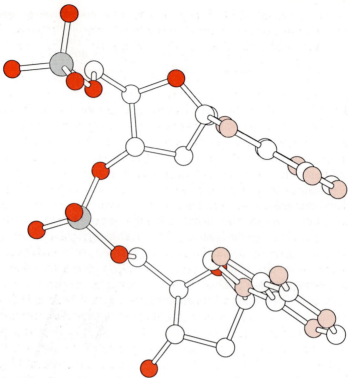

Figure 3.4 Conformation of two successive nucleotides in one strand of the A-DNA conformation, derived from fibre diffraction analysis.

pairs are twisted and tilted with respect to the helix axis (*Figure 3.4*), and are displaced nearly 5 Å from it, in striking contrast to the B helix. The helical rise is as a consequence much reduced, to 2.56 Å, compared with 3.4 Å for B-DNA. The helix is wider than the B one and has an 11 bp helical repeat (*Figure 3.3a*). The combination of base pair tilt with respect to the helix axis and base pair displacement from the axis results in very different groove characteristics for the A double helix compared with the B form. This also results in the centre of the A double helix being a hollow cylinder (*Figure 3.3d*). The major groove is now deep and narrow, while the minor one is wide and very shallow.

2.2 DNA polymorphism in fibres

The structural transition from A to B forms in 'mixed-sequence' DNA is induced by changes in the relative humidity surrounding the DNA fibres. Other forms can be produced by fine control of this condition (the C form) or from defined sequence polynucleotides (the D and Z forms). For example, there are a number of members of both the A and B families that differ from the 'canonical' forms outlined above, in large part because of the degree of over-winding of the helices (for the B-type family), and have considerable differences in pitch, base pair tilt with respect to the helix axis, and hence groove characteristics. For example, there are variants on the A-DNA duplex that have a wide major groove (4), reminiscent of that in a B helix, rather than that of the narrow standard A groove. Changes in groove width can thus be achieved at little energy cost.

The C form of DNA results from relatively low humidity conditions for a DNA fibre and is over-wound relative to B-DNA, with 9.3 residues per turn. The overall appearance of this helix and the dimensions of the grooves resemble those of the B-DNA rather than the A form. The D form cannot be adopted by random sequence native DNAs, and has been observed in crystalline fibres of the alternating purine/pyrimidine polynucleotides poly(dA–dT)·poly(dA–dT) and poly(dI–dT)·poly(dI–dT), where I is inosine (this is the rare 8-oxo-purine). The structure of this polymorph has not been unambiguously defined as yet, with models proposed ranging from left-handed seven- and eight-fold helices to a more conventional eight-fold right-handed one. Structural parameters for this right-handed model are given in *Table 3.1* and the Appendix. Observations of the reversible structural transition from B to D forms in crystalline fibres have been made using time-resolved X-ray diffraction by means of a high-intensity synchrotron source of X-rays (5). The observation of a gradual rather than an abrupt increase in helical pitch from 24 to 34 Å is consistent only with a transition in which there is no change in helix handedness, and so a left-handed D-DNA model can be rejected. The Z form is an authenticated left-handed structure (6). It exists in the alternating sequence poly (dC–dG)·poly(dC–dG) and is presumed to be the structure formed by this sequence in solution in high salt (> 2.5 M NaCl) conditions. This left-handed DNA has a dinucleotide repeat (*Figure 3.3c* and *f*) with quite distinct nucleoside conformations for the guanosine compared with the cytosine residues. Z-DNA is discussed further in Section 5 of this chapter; it was discovered in a defined-sequence oligonucleotide crystal structure prior to fibre diffraction analysis.

The polymorphism of DNA structure that is apparent in polymeric fibres is suggestive of an underlying flexibility in DNA structure. This is a natural consequence of the large number of backbone and sugar conformational variables, which together with base pair flexibility, can result in structurally distinct helices that are equal in energy. Fibre-diffraction studies can only hint at their fine detail, and so further significant advances in DNA structure studies only occurred with the advent of single-crystal analyses of defined oligonucleotide sequences. These not only provide much detail of the structures themselves, but also are gradually enabling the rules relating DNA structural features to sequence and environment to be evaluated.

3. B-DNA oligonucleotide structure as seen in crystallographic analyses

3.1 The Dickerson–Drew dodecamer

The single-crystal structure of the self-complementary dodecanucleotide d(CGCGAATTCGCG)$_2$ was determined in 1979 by multiple isomorphous re-

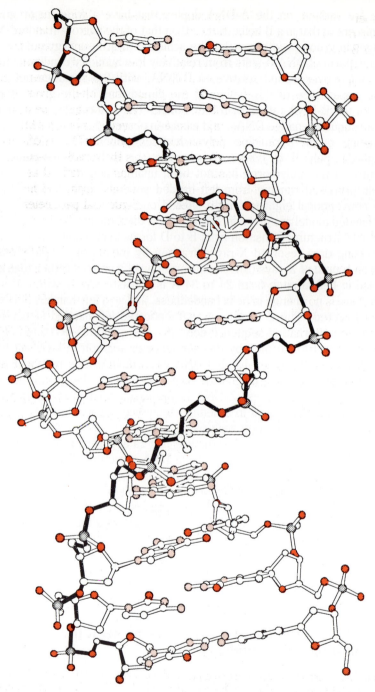

Figure 3.5. A view of the crystal structure of the B-DNA sequence d(CGCGAATTCGCG)$_2$ (refs. 7 and 8). The bonds for one backbone are shaded in black.

placement methods (7). This structure analysis is of landmark importance in that without recourse to any preconceived model, it showed an anti-parallel right-handed B-DNA double helix, and thus unequivocally demonstrated the correctness of the Watson–Crick model. This, the 'Dickerson–Drew' sequence, has subsequently been widely studied by a variety of other experimental and theoretical techniques which have confirmed and complemented the structural results. The crystallographic analysis, at 1.9 Å resolution, also revealed a number of major sequence-dependent structural features (*Figure 3.5*) that could not be observed in fibre-diffraction studies of averaged-sequence B-DNA, which has all the nucleotides in an identical conformation. These features are as follows.

1. There is a narrow minor groove in the 5'-AATT region, which at its extreme point is only 3.2 Å wide compared with 6.0 Å in fibre diffraction averaged B-DNA. The major groove at this point in the dodecamer structure is exceptionally wide (12.7 Å).

2. There is a well-ordered and regular network of water molecules in this region of the minor groove (8). This has been termed the 'spine of hydration'. It involves hydrogen bonding of first-shell water molecules to the O2 atom of thymine and the N3 atom of adenine such that each water molecule spans the two decamer strands (*Figure 3.6*). These water molecules are linked together by further, second-shell waters. Evidence for the persistence of this structured water arrangement in solution has recently come from NMR studies (9).

3. There are differences in a number of local base pair and base step parameters at different points along the sequence (10).

4. The values for sugar puckers, glycosidic angles, and backbone conformational angles have a wide distribution (*Figure 3.7*). The value for a particular parameter

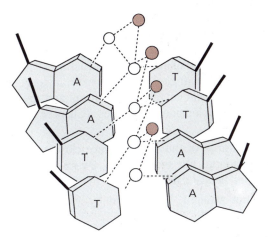

Figure 3.6.   Diagrammatic representation of the spine of hydration in the 'Dickerson–Drew' crystal structure of the sequence d(CGCGAATTCGCG)$_2$. Dashed lines indicate hydrogen bonds. First- and second-shell water molecules are shown as open and filled circles, respectively.

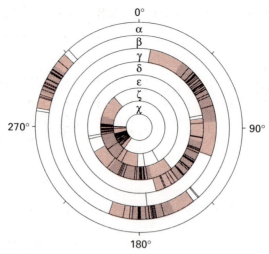

Figure 3.7. The distribution of values for the backbone torsion angles in the 'Dickerson–Drew' crystal structure (supplied by the Nucleic Acid Database). The angles are shown in a set of nested circles (conformational wheels), with individual angle values as straight lines in a particular wheel.

is to a considerable degree dependent on the sequence context of the nucleotide involved. Torsion angles ε and ζ have been found to be distributed in two groups: they are normally in the *trans, gauche⁻* domain, but are *gauche⁻, trans* when the nucleotide is followed by a purine. This alternative phosphate conformation has been designated B_{II}, with the standard *trans, gauche⁻* phosphate conformation being termed B_I. The B_{II} conformation is associated with increased minor groove width, both in the dodecamer and in a number of oligonucleotide crystal structures that have been subsequently determined. A number of other correlations between backbone torsion angles have been found, that extend to other, more recently determined oligonucleotide structures, for example between angles α and γ, and between angle δ and glycosidic angle χ. The δ angle is itself sugar pucker dependent; this correlation confirms and extends earlier findings from surveys of mononucleoside structures (Chapter 2, Section 5), to the oligonucleotide level.

5. The average features of the structure are close to those for canonical B-DNA from fibre diffraction. For example, the average helical twist is 35.9° and there are 10.1 bp per helical turn.

6. The helix is not straight, but is bent by about 19° in the major groove direction.

3.2 Sequence-dependent features

Variations in the local base pair and step parameters propeller twist, roll, and helical twist (see the Appendix for definitions) for this and several other dodecamer

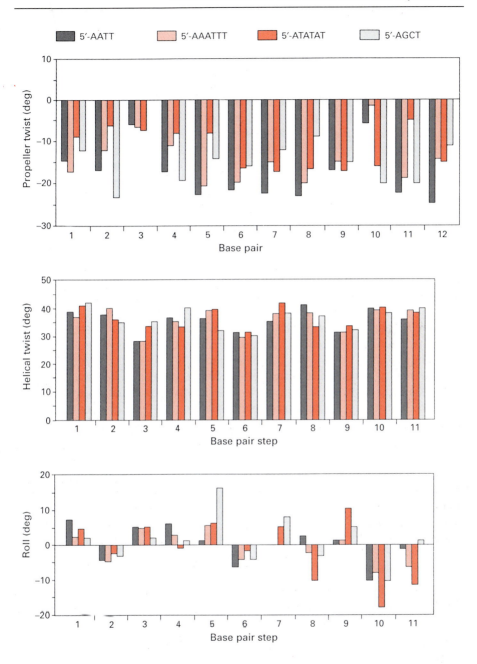

Figure 3.8. Plots of the variations in several local base parameters, for some dode-camer crystal structures: d(CGCGAATTCGCG)$_2$ (■) (ref. 7), d(CGCAAATTTGCG)$_2$ (■) (ref. 19), d(CGCATATATGCG)$_2$ (■) (ref. 16), and d(CGCAAGCTGGCG)$_2$ (□) (ref. 20). In each case, the horizontal axis indicates the base or base pair numbers along the sequence, from the 5′ end.

structures are illustrated in *Figure 3.8*. The sequence-dependent effects are most reliable for the central 6–8 residues of these sequences since the 3' and 5' termini of the dodecamer structures are involved in packing interactions with other molecules in the crystal. The most significant variations are as follows.

1. The local helical twist varies by up to 15°, with pyrimidine-3',5'-purine steps having lower than average values and purine-3',5'-pyrimidine steps having higher than average ones. These differences correlate with differences in the susceptibility of the sugar–phosphate backbone bonds in this sequence to be cleaved by the DNase I enzyme (11), with the TpC step, which has a high twist, being the most easily cleaved. This pattern of values is also consistent with the helical twists derived from NOE NMR data in solution (12). The mean twist angle in the crystal structure is 36°, in accord with the fibre diffraction value.

2. Propeller twists are significantly greater for A·T base pairs than for G·C ones, by an average of 5–7°.

3. Roll angles for pyrimidine-3',5'-purine steps have positive values, since they open up toward the minor groove, whereas purine-3',5'-pyrimidine steps have negative roll angles with major groove opening.

These sequence-dependent changes in helical and propeller twist, roll, and slide have been rationalized on the basis of steric clashes between substituent atoms on individual bases—the Calladine rules (13). The rules are based on the premise that the observed structural effects derive from opposite-strand purine–purine steric hindrances, and thus give an essentially mechanical view of DNA structural features as arising from the interplay of rigid-body Newtonian forces. These clashes are between major-groove substituent atoms O6 of guanine and N6 of adenine for purine-3',5'-pyrimidine steps, and between minor-groove N2 of guanine and N3 of adenine for pyrimidine-3',5'-purine steps. On this basis, little steric clash would be predicted for purine-3',5'-purine and pyrimidine-3',5'-pyrimidine steps. The rules suggest that clashes are avoided by combinations of changes in twist, roll, and slide, whose relative proportions depend on the nature of the bases involved:

(1) a decrease in local helical twist decreases minor-groove clashes;

(2) an increase in roll angle in the groove that has the clashes;

(3) separation apart of purines by means of an increase in slide;

(4) a decrease in propeller twist. The changes in slide result in alterations in the backbone angle δ, which thus has a higher value for purine than for pyrimidine nucleosides.

The Calladine rules, which have reasonable predictive capability for several B-DNA sequences other than d(CGCGAATTCGCG)₂, only have limited applicability to other helical types, especially A forms where there is inherently high propeller twist. The inability of the rules to take factors such as hydration, electrostatics, and hydrogen bonding into account, does mean that they only provide a first-order approximation to the understanding of sequence-dependent effects, even in B-DNA structures.

There is good evidence that the larger scale sequence-dependent groove width features seen in the dodecamer structure have direct relevance to 'real' DNA sequences and are not restricted to short crystalline ones. For example, it has been found (14) that the variations in the ability of the DNase I enzyme to cleave the phosphodiester backbone of the 160 bp *Escherichia coli tyr*T promoter within runs of AT base pairs relates to changes in minor groove width that occur in them when TpA steps are present or absent.

3.3 Other B-DNA oligonucleotide structures

The Calladine rules provide a means of understanding the structural behaviour of dinucleotide base steps, of which there are just 10 distinct types. The rules imply that each step is largely independent of its neighbours, a considerable over-simplification. Extension to four-base sequences would require information for 136 distinct possible combinations. The establishment of the first B-DNA oligo-nucleotide structure provided information on only a very small number of these possible nearest-neighbour combinations of DNA sequence, and so there have been concerted attempts to determine structures for other sequences, both relating to and distinct from the Dickerson–Drew one. It has proved remarkably difficult to crystallize more than a small range of sequences, and as yet it has not been possible to derive second-generation Calladine-like rules of general validity since there are structural data on only approximately a quarter of the 136 four-base steps.

The crystal structures of several dozen standard B-type oligonucleotides have now been reported, all of which are anti-parallel double helices. Their averaged features closely approximate those of averaged-sequence fibre-diffraction B-DNA (detailed in the Appendix). None is more than 12 bp in length. Almost all are self-complementary palindromic sequences. They fall into two principal categories. They are either dodecamers, the overwhelming majority of which are isomorphous to the Dickerson–Drew structure and therefore pack identically in the crystal, in the space group $P2_12_12_1$ or they are decamers, which crystallize in a variety of space groups, some of which give very high resolution (1.3–1.4 Å) structures. The isomorphism of the dodecamers provides a convenient frame-work to analyse systematic changes in the central base pairs, whilst keeping invariant the 3' and 5' terminal sequences, which are involved in the crystal packing arrangements (15). Several decamers crystallize in forms with the 10-mer duplexes packed end-to-end in the crystal, in a quasi-helical manner.

A number of changes to the central AT sequence of the dodecamer have been examined. When the sequence is a purely alternating one, as in d(CGCATATATGCG)$_2$ (16), the minor groove width is equally narrow as in the Dickerson–Drew structure, although the narrowest points do not coincide. This ATATAT sequence shows a pronounced alternation of several of its base pair and step parameters, especially roll; propeller twist values for the A·T base pairs are consistently lower than those in the Dickerson–Drew sequence. The

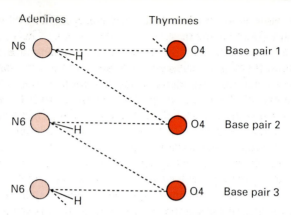

Figure 3.9. Schematic diagram of the arrangement of bifurcated hydrogen bonds (shown as dashed lines) in the major groove of an A tract, for three consecutive A·T base pairs, between donor N6 and acceptor O4 atoms.

structure of the duplex formed by d(CGCAAAAAAGCG) and its complementary sequence (17) is remarkable for several reasons. It is the closest model sequence yet studied to an A tract, which in natural DNA has anomalous structural and dynamic properties—it cannot be readily reconstituted into nucleosomes and is intrinsically bent when in phase with a helical repeat. The structure has high propeller twists for the A·T base pairs, of up to 26°, with a network of bifurcated three-centre hydrogen bonds connecting them (*Figure 3.9*). These bifurcated hydrogen bonds are in the major groove and involve atoms N6 of adenines and O4 of thymines. They are presumed to be major factors in stabilizing and stiffening the A tract. The roll, tilt, and slide values for the ApT steps in the structure are all close to 0°, with bending (towards the major groove) occurring at the ends of the A tract rather than within it. Bending in the major groove direction has also been observed in the structure of the duplex formed by d(CGCAAAAATGCG) and its complementary strand (18), although the situation here is complicated by the presence of two distinct orientations of the duplex in the crystal lattice. This study has concluded that the bending found in the isomorphous dodecamer crystal structures is more a consequence of crystal packing forces than of intrinsic properties of the A tract sequences. The high (~20°) propeller twists for A·T base pairs found in the structure of d(CGCAAATTTGCG)$_2$, are not accompanied by significant three-centre hydrogen bonding in the AT region, suggesting (19) that these effects may not be required to stabilize high propeller twist and a narrow minor groove. Rather, propeller twist is more a consequence of base stacking preferences.

Further confirmation for the strong tendency of only AT regions to have a narrow minor groove has come from the analysis of the duplex formed by d(CGCAAGCTGGCG)$_2$, which has an average minor groove width of about 7 Å in the AGCT region, which is close to that of averaged fibre diffraction B-DNA (20), as does the GC-rich minor groove in the d(ACCGGCGCCACA) duplex (21). This latter non-palindromic sequence is a hot-spot for frameshift mutagenesis by several chemical carcinogens. The structure is non-isomorphous with the other dodecamers and is remarkable for several features. Three G·C base pairs in the centre of the helix are partially opened, with exceptionally high propeller twists and some loss of G·C hydrogen bonding. It has been suggested that this represents the intermediate stage prior to full base pair opening. Analysis of a low temperature form of the crystals has shown (22) that cooling produces a 1 bp shift of the major groove base pairing that is favoured by CA tracts. The crystal packing involves helices interacting together by means of groove–backbone interactions, which may be a good model for DNA–DNA junctions such as the Holliday junction involved in recombinational processes (23).

Recent crystallographic analyses of a number of decamer helices (see for example, references 24–27) has considerably extended knowledge of the variations in sequence-dependent structural features. Comparisons of the structures show that the parameters helical twist, roll, rise, and cup are not independent of each other, but that there are several correlations between them. Rise is linearly related to helical twist, and is also a linear function of the difference between roll and cup:

$$\text{Rise} = 3.377 + 0.038 \, (\text{roll} - \text{cup})$$

A 'Profile Sum' (PS) has been defined (25) that relates this and other linear correlations of these four parameters, whose coefficients arise from the correlation analyses:

$$\text{PS} = (\text{twist} - 36) - 16.24(\text{rise} - 3.36) + 0.744 \times \text{cup} - 0.703 \times \text{roll}$$

Some rules governing base step behaviour have been proposed that are based on PS values. High twist profile steps (HTP) have high helical twist, low rise, high cup, and low roll, whereas low twist profile steps (LTP) have low helical twist, high rise, low cup, and high roll. Other steps have intermediate twist profile (ITP) and a few show variable twist profile (VTP), highly dependent on the nature of the sequences flanking the step. Typical HTP values are +10 to +20, while LTP values range from −10 to−20. Base steps in the known decamer duplexes can be classified on this basis as:

> HTP: CpA, GpA, GpC
> LTP: ApG, ApA, GpG
> ITP: ApT, ApC
> VTP: CpG, TpA

Thus, all purine-3′,5′-purine steps are LTP except for GpA, and most purine-3′,5′-pyrimidine steps, except for GpC, are ITP. The TpA step, even though it is consistently found to be unstacked in all the B-DNA oligonucleotide structures where it occurs, has in terms of profile sum highly variable sequence and crystal environment-dependent behaviour. An extreme and surprising example of this is in the sequence d(CGATATATCG)$_2$, which has a wide minor groove in the alternating AT region (26), albeit with a characteristic alternation of helical twist angles that is a consequence of the alternation of good base stacking at ApT steps combined with the poor stacking at the TpA ones. Thus, although alternating AT sequences have a marked tendency to produce narrow minor grooves in B-DNA, their inherent structural flexibility means that this tendency can be readily overcome under some circumstances. For example, the B$_{II}$ phosphate conformation, which occurs only at steps with the second base being a purine, is itself a major factor in producing a widened minor groove.

4. A-DNA oligonucleotide crystal structures

DNA double helices of the A type are produced in fibres of random-sequence DNA under conditions of low humidity (see Section 2 above), and in solution when the water activity is reduced by the addition of various alcohols. The fibre diffraction studies suggest that certain runs of sequence, such as alternating G·C/C·G base pairs, may have a tendency to be in an A form. Single-crystal analyses of particular lengths of oligonucleotide sequences have determined many to be of the A-type. However, for the majority of such structures, it is now apparent that this reflects not so much any inherent structural preference, but rather the crystal packing requirements for these lengths of oligonucleotide duplexes. The high-alcohol crystallization conditions commonly used in oligonucleotide crystallography may also be a contributory factor. None the less, these structures have provided invaluable insights into the flexibility and conformational preferences of A-DNA helices in general, even if their significance in biological terms is less clear. There is one instance where A-type structures are of undoubted importance: this is in the case of RNA–DNA hybrids, formed, for example, during transcription and replication. Since RNA helices are always constrained to be in an A form (on account of the stereochemical constraints imposed by the 3′ hydroxyl group), then DNA–RNA hybrids would be expected to be similarly constrained. This has been found to be the case both by fibre diffraction studies of RNA–DNA hybrid polynucleotides and by single-crystal analyses. The structure of the duplex formed by r(GCG)d(TATACGC) shows that the DNA strand has adopted a conformation close to that of duplex RNA with an 11-fold helix (average helical twist of 33°), C3′*endo* sugar puckers, and an A helical type backbone conformation (28).

4.1 A-form octanucleotides

A number (>20) of self-complementary octanucleotides, of widely varying sequence type, have been crystallized and found to have A-like structures

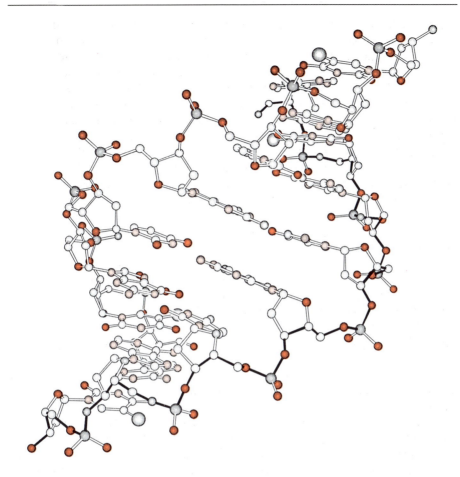

Figure 3.10.   View of a typical A-DNA octanucleotide crystal structure, for the sequence d(ATGCGCAT)$_2$ (ref. 34).

(*Figures 3.10*). As with B-type oligomers, structural features averaged over a number of such sequences are close to those of fibre diffraction A-DNA (see the tables in the appendix). Some of these sequences crystallize in the tetragonal space group, P4$_3$2$_1$2; others are in the hexagonal one, P6$_1$. It has been possible to vary conditions such that some sequences such as d(GGGCGCCC)$_2$ (29) and d(GTGTACAC)$_2$ (30) have been crystallized in both forms. The hexagonal form has also been found for sequences such as d(GGGGCCCC)$_2$ (31), d(GGGTACCC)$_2$ (32), and d(GGGATCCC)$_2$ (33) as well as for a number of mismatch variants (see Section 4.1). All have helical and conformational features within the general A class, but with some marked local variations, in particular in the overall helical repeat, in minor-groove widths, and in base step parameters. The wide minor groove in d(GGGGCCCC)$_2$ (~15Å) contrasts with the more typical value of 9.7–9.8 Å in d(GGGTACCC)$_2$, which is close to that for canonical fibre-diffraction A-DNA (11.1 Å). Major groove width cannot be fully assessed in an octanucleotide duplex, since eight rather than seven phosphate groups are required in order to calculate the shortest phosphate–phosphate interstrand distance. Approximated major-groove widths show very wide variation, ranging from 5 Å in the low-temperature form of d(GGGCGCCC)$_2$ (29) to over 12 Å in d(GGGGCCCC)$_2$ (31), compared with the very narrow fibre value of 2.2 Å. The tetragonal form, with sequences having a pyrimidine-3′,5′-purine step at positions 4 and 5, that is at the centre of the helix, shows significant deviations from standard A-form backbone geometry at this point, with torsion angles α and γ having *transoid* values rather than the normal *gauche*$^-$ and *gauche*$^+$ ones respectively. This discontinuity has been observed in several structures, for example that of d(ATGCGCAT)$_2$ (34). It is likely that this effect is a consequence of the requirements imposed by hydration and crystal packing in the tetragonal unit cell, rather than being an intrinsic property of these sequences.

4.2 A-form oligonucleotides in solution? Crystal packing effects

Indeed, crystal packing factors may well be the driving force behind *all* octanucleotide structures being in the A form. Solution NMR studies on the sequence d(ATGCGCAT)$_2$ (34) have demonstrated that it shows B-family behaviour in solution even though the crystal structure is unequivocally of the A type. This dichotomy between crystal and solution environmental constraints is also shown by the sequence d(GGATGGGAG), which is a part of the binding site for the transcription factor TFIIIA, from the initiation complex for transcription of the 5S RNA gene in *Xenopus*. The (low-resolution) 3 Å crystal structure of the duplex formed by this sequence and its complementary strand, shows an A-family helix (35), with average major- and minor-groove widths of 14.8 and 15 Å respectively, and a helical repeat of 11.5 bp. These values closely correspond to those for the A′-RNA fibre-diffraction structure. Nuclease digestion studies, showing a repeat of 11.4 bp per turn, together with circular dichroism (CD) measurements (36), are consistent with the assignment of a structure for the d(GGATGGGAG) sequence as well as for the complete 54 bp TFIIIFA binding site, that is not classical B-DNA. On the other hand, NMR and other biophysical methods all indicate that the nonamer in solution has normal C2′*endo* sugar puckers and B-DNA range glycosidic angles (37). It should be borne in mind that NMR methods do not directly give information on long-range distances such as groove widths, so that one cannot exclude a model for the TFIIIA binding site that has some elements of both A- and B-DNA helix features, as has been found in several polynucleotide polymorphs.

Short A-DNA double helices are insufficient in length for the major groove to be fully formed, and so estimates of its dimensions are necessarily approximate. Recent X-ray analyses of two dodecamers (in two distinct space groups) have each revealed the structure of a complete turn of A-DNA double helix, as well as conformational features that are not subject to the crystal packing factors of the octanucleotides. The sequence d(CCCCCGCGGGGG)$_2$, with both alternating and non-alternating nucleotides (38) has average backbone conformational angles and helical parameters that are remarkably close to those in A-DNA fibres, with some local backbone angle variations, for example *trans* α and γ angles at one of the two CpG steps. This structure, together with that of the A-DNA decamer d(ACCGGCCGGT)$_2$ (39), is consistent with the notion that very GC-rich sequences have a higher tendency to form A-type duplexes. The sequence d(CCGTACGTACGG)$_2$, of which two-thirds consists of G·C base pairs, likewise has an A-DNA structure (40), this time without any *transoid* α/γ angles. The average major groove width in this structure, of 3.5 Å, is similar to the 2.2 Å fibre diffraction value.

5. Z-DNA—left-handed DNA

5.1 The DNA hexanucleotide crystal structure

The first oligonucleotide duplex single-crystal structure to be solved (by multiple isomorphous replacement methods) was that of the alternating pyrimidine–purine sequence d(CGCGCG)$_2$ in 1979, at the very high resolution of 0.9 Å (41). This structure is remarkable in that even though it consists of a duplex formed by the two anti-parallel hexamer strands, the helix is (quite unexpectedly) a left-handed one (*Figure 3.11*). The backbone is irregular compared with A- or B-DNA since the dC and dG residues have very distinct conformations (*Figure 3.12*), resulting in a 'zigzag' arrangement of phosphate groups; hence the helix has been termed Z-DNA. The same left-handed arrangement has also been found in fibres of the GC alternating polynucleotide (6). Z-DNA was in effect discovered some years earlier during the course of CD studies on poly(dG–dC)·poly(dG–dC), when it was observed (42) that on increasing salt concentration beyond ~4 M NaCl, the CD spectrum became inverted from its standard B-DNA-associated shape. This major spectral change indicates a transition to a quite different conformational form, which is now accepted to be Z-DNA. A number of other physico-chemical techniques have subsequently been used to study this left-handed structure in addition to X-ray techniques (43).

5.2 Structural features

Z-DNA oligonucleotides (and the poly(dG–dC)·poly(dG–dC) polynucleotide) have the following conformational characteristics:

1. The purine (deoxyguanosine) nucleosides have *syn* glycosidic angles with a χ range of 55–80° (mean 60°), together with C3′*endo* sugar puckers. The conformation about the α/γ backbone torsion angles is *gauche*$^+$, *trans*, and *gauche*$^-$ about angle ζ.

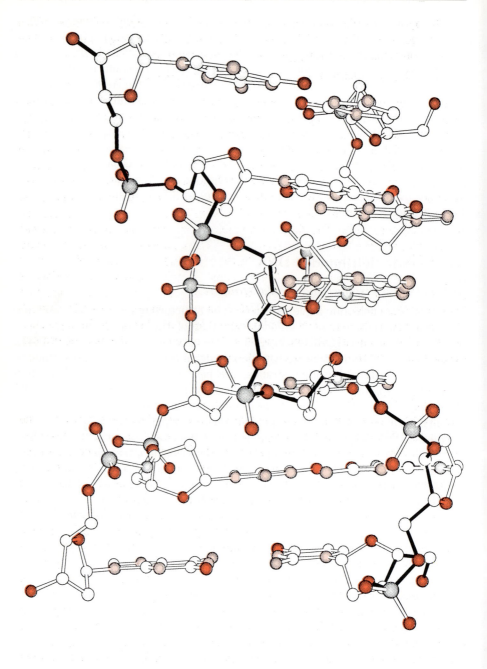

Figure 3.11.   A plot of the structure of Z-DNA as seen in the d(CGCGCG)$_2$ crystal structure (ref. 41). The bonds in one backbone have been shaded black to emphasize the left-handed nature of the duplex.

2. The pyrimidine (deoxycytidine) nucleosides have *anti* glycosidic angles with a χ range of -145 to $-160°$ (mean $-152°$) and C2'*endo* sugar puckers. The α/γ conformation is *trans, gauche$^+$* and that about angle ζ is *gauche$^+$*.

3. The G·C base pairs are of standard Watson–Crick type.

A consequence of these differences between purine and pyrimidine nucleosides is that the helical repeating unit is forced to be the CpG *di*nucleoside, rather than the *mono*nucleoside one in standard right-handed canonical A- and B-DNA. The distinct α/ζ phosphate conformations of the two nucleosides result in the characteristic zigzag appearance of the backbone. Detailed examination of d(CGCGCG)$_2$ in various crystal forms (44) has shown that there is a secondary backbone conformational family with a distinct set of phosphate orientations. This secondary conformation is termed Z_{II}, with the standard type as described above being termed Z_I. The Z_{II} conformation results in purines having a *gauche$^+$* rather than a *gauche$^-$* value for the ζ angle and a *gauche$^-$* rather than a *trans* value for angle ε. Z_{II} pyrimidines have (rather smaller) changes in angles α and β than do Z_I ones. Occurrence of the Z_{II} conformation at a particular residue is probably related to the coordination of the phosphate group at this point to a hydrated magnesium ion (44, 45): Z-DNA oligonucleotides are usually crystallized in the presence of magnesium and spermine ions. Thus, the Z_{II} conformation does not occur at the same point in the sequence in the pure-spermine, magnesium-free structure of d(CGCGCG)$_2$.

5.3 The Z-DNA helix

The Z-DNA double helix is to a large degree represented by the structure of the hexanucleotide duplex d(CGCGCG)$_2$; however, there are slight differences between individual CpG units, quite apart from the occurrence of Z_I and Z_{II} forms. Idealized helices for both of these have been generated (44). The fibre-diffraction model for Z-DNA is closest to the Z_I helix. This crystallographically derived Z helix has a 44.6 Å pitch with 12 bp per helical turn and a diameter of ~18 Å (*Figure 3.3c* and *f*), making it somewhat slimmer than a B-DNA helix (with a diameter of ~20 Å). The helical twists for 2 bp steps CpG and GpC are quite different (see the Appendix) as a consequence of the asymmetry in guanosine and cytidine conformations, with the CpG step having an exceptionally small twist and almost no base stacking between the two C·G base pairs in the step. A further consequence of the *syn* guanosine conformation is that the base pairs are not positioned astride the helix axis, as in B-DNA. Rather, their edges are at the surface of the helix. The N7 and C8 atoms of the imidazole 5 membered ring of guanine and (to a lesser extent) the C5 atom of cytosine, actually protrude on to the helical surface of what would be the major groove, making the surface slightly convex at this point (*Figure 3.3c*). In other words, Z-DNA does not have a major groove at all. Its minor groove is very narrow and deep, and lined with phosphate groups (*Table 3.1*). The differences in phosphate orientation between Z_I and Z_{II} helices result in the latter having a ~1 Å wider minor groove.

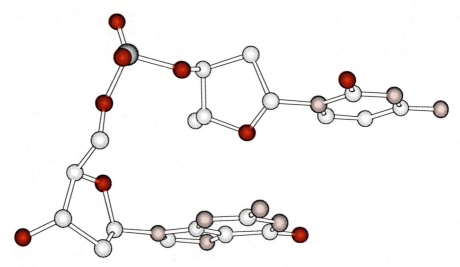

Figure 3.12. A detailed view of the GpC dinucleoside repeat in Z-DNA, from the d(CGCGCG)$_2$ crystal structure (ref. 41).

5.4 Other Z-DNA structures

Z-DNA oligonucleotides have been crystallized with a range of related sequences, including pure analogues such as d(CGCGCGCG)$_2$ (46), as well as variants where changes have been made to base and sequence type. All show the resilience of the Z-DNA structural entity. The central CpG can be replaced by TpA whilst maintaining the left-handed structure (47). This shows that Z-DNA can tolerate some A·T base pairs, although these do tend to destabilize the structure. The cytosines are required to be 5-methylated in order for the d(CGTACG)$_2$ sequence to be stabilized, and in general modified cytosines are necessary if a Z-DNA structure contains A·T base pairs. It has been suggested (47) that A·T base pairs are not able to take part in the ordered Z-DNA groove hydration that plays an important role in maintaining the integrity of the Z-DNA structure (45), by contrast with G·C base pairs. Hence, the former cannot by themselves form a Z-DNA structure. The slight preference of guanosine nucleosides (unlike adenosine) to adopt the *syn* rather than the *anti* glycosidic conformation (see Chapter 2, Section 4) is also a factor in the GC preference of Z-DNA. It is then surprising that reversal of the central pyrimidine-3′,5′-purine sequence (i.e. so that it is no longer purely alternating) in the structure of d(CGATCG)$_2$, still retains a Z-DNA type structure (48), albeit with the thymidines and adenines adopting *syn* and *anti* glycosidic conformations respectively. This energetically unfavourable Z structure was only stabilized by having the cytosines methylated or brominated at the 5 position of the base. None the less, its existence suggest that the requirement for the formation of a Z-DNA structure is not so much that the sequence should be an alternating CG-rich one, but that it has alternating *syn* and *anti* nucleosides. It is likely that replacement of thymine with uracil would increase Z-form stability, since the methyl group of the former presents a significant hydrophobic hindrance to solvent ordering around the Z helix. This has been borne out by subsequent structural studies on uracil-containing Z sequences (49).

5.5 Biological aspects of Z-DNA

The accessibility of the C8 (and to a lesser extent the N7) position of guanine in Z-DNA renders it susceptible to attack by several electrophilic compounds that have a preference for these positions. Examples include the *N*-aryl carcinogens acetylaminofluorene and aflotoxin, as well as C8-bromination. All stabilize Z-DNA and promote the B to Z structural transition. Thus, Z-DNA tracts within genomic sequences could act as mutational hot-spots for these agents. The fact that Z-formation is also facilitated by methylation at the C5 position of cytosine, a possible mechanism in some eukaryotic organisms for regulating gene transcription, suggests that Z-DNA could play a role in gene regulation.

Z-DNA in its linear form is less stable than either the A or B forms, requiring high salt or alcohol concentrations for maintenance of its structure in solution. These conditions reduce electrostatic interactions between interstrand phosphate groups, which in Z-DNA are much closer together than in B-DNA (7.7 Å in the Z_I form compared with 11.7 Å across the minor groove in B-DNA). Z-DNA sequences can also be stabilized at physiological salt levels when incorporated in underwound, negatively supercoiled, covalently-closed circular DNA, for example in plasmid DNA. Even low (~1–2 per cent) levels of Z-forming sequence can significantly change supercoiling properties. It has been proposed (50) that a role for Z-DNA *in vivo* is to act as a signal for the induction of transcription via supercoiling.

The classic Z-forming sequence, $d(CG)_n$, is much rarer in eukaryotic than in prokaryotic organisms. However, the sequence $d(CA/GT)_n$, which can also form left-handed DNA in negatively supercoiled plasmids, is much more prevalent in eukaryotic genomes. The overriding question of whether Z-DNA actually exists naturally, rather than whether it can be formed under appropriate circumstances, has not as yet been fully answered. A number of proteins that bind to Z-DNA have been isolated, although their role(s) and significance remain unclear.

6. Bent DNA

6.1 DNA periodicity in solution

DNA under physiological solution conditions is generally assumed to be in a B-type conformation, even when combined with chromosomal proteins. There are at first sight some important differences between the B structures as found in the crystal and fibre, and that in solution, even as naked DNA. The structural studies all point to a helical repeat of 10.0–10.1 residues per turn for averaged-sequence DNA. However, plasmid mobility, nuclease digestion, and hydroxyl radical cleavage experiments (51) have given a rather different value, of close to 10.5 bp per turn, again averaged over many base pairs. This apparent discrepancy is only partly resolved by the observation that helical periodicity in solution is markedly sequence dependent (52), and that cleavage at a particular point along a DNA sequence depends on the local helical twist and hence groove width, in a manner that fully correlates with the crystallographically determined features. It may well be that the constraints of crystallization in a particular lattice (especially when the helical oligonucleotide structures pack in an end-to-end manner) impose particular values on helical twist that are not always representative of a given oligonucleotide sequence when in solution. It is reassuring, though, that most other DNA conformational and base pair parameters are essentially unchanged on going from solution to the crystalline environment.

6.2 A tracts and bending

The DNA double helix is usually thought of as a straight molecule. The original Watson–Crick model is of a straight helix, since it was based on data from 'straight' DNA fibres. In fact, it must be bent or curved in many of its functional roles, and is deformed from linearity in the dodecamer crystal structures, being bent by about 19°. There has been much speculation on the nature of DNA bending in such situations as chromatin (which has ~145 bp wound around a histone protein core) and closed circular DNA. Some sequences have an inherent tendency to bend independently of there being any protein present. The best-studied such sequences are those comprising runs of adenines on one strand and thymines on the other (termed A tracts).

Intrinsic DNA bending and curvature were discovered (53) in restriction fragments of kinetoplast DNA, which have extensive repeats of A tracts within them, each having about 5–7 bp. These fragments showed highly anomalous gel electrophoresis behaviour, with very slow migration through the gel, as if the DNA had a much higher molecular weight than was the case. Bending occurs when sequences or motifs such as 5'-AAAAAA are repeated in phase with the DNA helical repeat itself (54, 55), that is every 10 or 11 bp. The phasing is critical since it forces the tract to be always on the same side of the helix, thereby enabling the individual small bends associated with each tract to add up to a significant amount of total bending. It has been estimated (56) that an individual A_6 tract bends a helix by about 17–21°. A typical bending sequence, from kinetoplast DNA, is:

5'-.....CCAAAAATGTCAAAAAATAGGCAAAAAATGCCAAAAAT.....

Two distinct structural models have been suggested to account for these observations of bending. In one, a combination of gel retardation and NMR studies has been used (57), resulting in a structural model that has bending towards the minor groove direction of the helix. The compression and closure of the minor groove is in part a result of negative A·T base pair tilt within the A tract itself. In this, the 'junction' model, bending is caused by the abrupt change in structure when the A tract meets the more normal B-DNA intervening sequence, resulting in the helix axes of the two segments being at an angle to each other. Abrupt changes in either tilt or roll can produce this situation. Thus, in the junction model, the A tracts are themselves not bent. By contrast, the 'wedge' model (58) has continuous bending along the A tract as a result of changes in roll at each dinucleotide step—this model has also been suggested for the continuous (induced) bending of general-sequence DNA when wrapped around histones in nucleosomes.

The crystal structure of the decamer d(CATGGCCATG)$_2$ shows features relevant to the general curvature issue. The helix is not straight but has a smooth 23° bend over the four central base pairs (59). This intrinsic bend is only consistent with a straight A tract model, although the roll compression of the major groove suggests a third type of model.

6.3 The structure of poly- and oligo(dA·dT)

Several oligonucleotide crystal structures contain short dA·dT sequences that might be thought of as models for A tracts, as discussed in Sections 3.1 and 3.2 above. These have narrow minor grooves with ordered water molecules, high propeller twists for the A·T base pairs, together with, in two cases (17, 18), bifurcated three-centre hydrogen bonds involving adjacent A·T base pairs (*Figure 3.10*). Such hydrogen bonding was, however, not observed (19) in the somewhat higher resolution structure of d(CGCAAATTTGCG)$_2$. The A$_6$·T$_6$ tract in one crystal structure (17) is itself little bent, lending some support to the junction model. All three crystal structures have an overall bending towards the major groove direction, in contrast to the inference from gel studies that A tracts are bent towards the minor groove. Once again, crystal packing forces have been invoked to account for this discrepancy. It is more likely that the structures, since they do not fulfil the sequence requirements for A tract phasing, are simply inadequate as realistic models to account for all the features of A tracts.

There is increasing evidence that A tracts do not inherently require three-centre A·T hydrogen bonds for curvature to be produced. For example, sequences having inosine·cytosine base pairs (these cannot form major groove three-centre hydrogen bonds since inosine lacks an N6 group), still result in bent structures (60).

The structure of poly(dA·dT) itself has been the subject of a number of studies. Some have used fibre diffraction methods to define an experimental model, with the 'heteronomous' model (61) having distinct backbone conformations and sugar puckers for the two dA and dT strands. The results of more recent analyses (62, 63), although differing in detail, support a structure with both strands having B-like conformations, albeit with a narrow minor groove. Propeller twists for the A·T base pairs are likely to be significant (~15°), but not as high as in the earlier crystal structures, thus reducing the possibility of three-centre hydrogen bonding. Molecular dynamics studies (64) similarly indicate that this type of hydrogen bonding is not of primary importance in stabilizing oligo- or poly(dA·dT), but merely that it can occur when the geometric circumstances allow.

7. Further reading

Arnott,S. (1970). *Progress in Biophysics and Molecular Biology*, **6**, 265.
Dickerson,R.E. (1992). *Methods in Enzymology*, **211**, 67.
Dickerson,R.E., Grzeskowiak,K., Kopka,M.L., Larsen,T., Lipanov,A., Privé,G.G., Quintana,J., Schultze,P., Yanagi,K., Yuan,H., and Yoon,H.-C. (1991). *Nucleosides & Nucleotides*, **10**, 3.
Drew,H.R., McCall,M.J., and Calladine,C.R. (1988). *Annual Review of Cell Biology*, **4**, 1.
Kennard,O. and Hunter,W.N. (1989). *Quarterly Reviews of Biophysics*, **22**, 3.
Shakked,Z. and Rabinovich,D. (1986). *Progress in Biophysics and Molecular Biology*, **47**, 159.

References

1. Dickerson,R.E. *et al.* (1989). *The EMBO Journal*, **8**, 1.
2. Watson,J.D. and Crick,F.H.C. (1953). *Nature*, **171**, 737.
3. Chandrasekaran,R. and Arnott,S. (1989). In *Landolt-Börnstein, New Series*, Group VII, Vol. 1b (ed. W.Saenger). Springer-Verlag, Berlin.
4. Arnott,S., Chandrasekaran,R., Hall,I.H., Puigjaner,L.C., Walker,J.K., and Wang,M. (1982). *Cold Spring Harbor Symposia on Quantitative Biology*, **47**, 53.
5. Mahendrasingam,A., Forsyth,V.T., Hussain,R., Greenall,R.J., Pigram,W.J., and Fuller,W. (1986). *Science*, **233**, 133.
6. Arnott,S., Chandrasekaran,R., Birdsall,D.L., Leslie,A.G.W., and Ratliffe,R.L. (1980). *Nature*, **283**, 743.
7. Wing,R.M., Drew,H.R., Takano,T., Broka,C., Takana,S., Itakura,K., and Dickerson,R.E. (1980). *Nature*, **287**, 755.
8. Drew,H.R. and Dickerson,R.E. (1981). *Journal of Molecular Biology*, **151**, 535.
9. Kubinec,M.G. and Wemmer,D.E. (1992). *Journal of the American Chemical Society*, **114**, 8739.
10. Dickerson,R.E. and Drew,H.R. (1981). *Journal of Molecular Biology*, **149**, 761.
11. Lomonossoff,G.P., Butler, P.J.G., and Klug,A. (1981). *Journal of Molecular Biology*, **149**, 745.
12. Lane,A.N., Jenkins,T.C., Brown,T., and Neidle,S. (1991). *Biochemistry*, **30**, 1372.
13. Calladine,C.R. and Drew,H.R. (1984). *Journal of Molecular Biology*, **178**, 773.
14. Drew,H.R. and Travers,A.A. (1984). *Cell*, **37**, 491.
15. Dickerson,R.E., Goodsell,D.S., Kopka,M.L., and Pjura,P.E. (1987). *Journal of Biomolecular Structure and Dynamics*, **5**, 557.
16. Yoon,C., Privé,G.G., Goodsell,D.S., and Dickerson,R.E. (1988). *Proceedings of the National Academy of Sciences, USA*, **85**, 6332.
17. Nelson,H.C.M., Finch,J.T., Luisi,B.F., and Klug,A. (1987). *Nature*, **330**, 221.
18. DiGabriele,A.D., Sanderson,M.R., and Steitz,T.A. (1989). *Proceedings of the National Academy of Sciences, USA*, **86**, 1816.
19. Edwards,K.J., Brown,D.G., Spink,N., Skelly,J.V., and Neidle,S. (1992). *Journal of Molecular Biology*, **226**, 1161.
20. Webster,G.D., Sanderson,M.R., Skelly,J.V., Neidle,S., Swann,P.F., Li,B.F., and Tickle,I.J. (1990). *Proceedings of the National Academy of Sciences, USA*, **87**, 6693.
21. Timsit,Y., Westhof,E., Fuchs,R.P.P., and Moras,D. (1989). *Nature*, **341**, 459.
22. Timsit,Y., Vilbois,E., and Moras,D. (1991). *Nature*, **354**, 167.
23. Timsit,Y. and Moras,D. (1991). *Journal of Molecular Biology*, **221**, 919.
24. Quintana,J., Grzeskowiak,K., Yanagi,K., and Dickerson,R.E. (1992). *Journal of Molecular Biology*, **225**, 379.

25. Yanagi, K., Privé, G. G., and Dickerson, R. E. (1991). *Journal of Molecular Biology*, **217**, 201.
26. Yuan, H., Quintana, J., and Dickerson, R. E. (1992). *Biochemistry*, **31**, 8009.
27. Heinemann, U., Alings, C., and Bansal, M. (1992). *The EMBO Journal*, **11**, 1931.
28. Wang, A. H.-J., Fujii, A., van Boom, J. H., van der Marel, G. A., van Boeckel, S. A. A., and Rich, A. (1982). *Nature*, **299**, 601.
29. Shakked, Z., Guerstein-Guzikevich, G., Eisenstein, M., Frolow, F., and Rabinovich, D. (1989). *Nature*, **342**, 456.
30. Jain, S., Zon, G., and Sundaralingam, M. (1991). *Biochemistry*, **30**, 3567.
31. McCall, M., Brown, T., and Kennard, O. (1985). *Journal of Molecular Biology*, **183**, 385.
32. Eisenstein, M., Frolow, F., Shakked, Z., and Rabinovich, D. (1990). *Nucleic Acids Research*, **18**, 3185.
33. Lauble, H., Frank, R., Blöcker, H., and Heinemann, U. (1988). *Nucleic Acids Research*, **16**, 7799.
34. Clark, G. R., Brown, D. G., Sanderson, M. R., Chwalinski, T., Neidle, S., Veal, J. M., Jones, R. L., Wilson, W. D., Zon, G., Garman, E., and Stuart, D. I. (1990). *Nucleic Acids Research*, **18**, 5521.
35. McCall, M., Brown, T., Hunter, W. N., and Kennard, O. (1986). *Nature*, **322**, 661.
36. Fairall, L., Martin, S., and Rhodes, D. (1989). *The EMBO Journal*, **8**, 1809.
37. Aboul-ela, F., Varani, G., Walker, G. T., and Tinoco, I, Jr (1988). *Nucleic Acids Research*, **16**, 3559.
38. Verdaguer, N., Aymami, J., Fernández-Forner, D., Fita, I., Coll, M., Huynh-Dinh, T., Igolen, J., and Subirana, J. A. (1991). *Journal of Molecular Biology*, **221**, 623.
39. Frederick, C. A., Quigley, G. J., Teng, M.-K., Coll, M., van der Marel, G., van Boom, J. H., Rich, A., and Wang, A. H.-J. (1989). *European Journal of Biochemistry*, **181**, 295.
40. Bingman, C. A., Zon, G., and Sundaralingam, M. (1992). *Journal of Molecular Biology*, **227**, 738.
41. Wang, A. H.-J., Quigley, G. J., Kolpak, F. J., Crawford, J. L., van Boom, J. H., van der Marel, G. and Rich, A. (1979). *Nature*, **282**, 680.
42. Pohl, F. M. and Jovin, T. M. (1972). *Journal of Molecular Biology*, **67**, 375.
43. Rich, A., Nordheim, A., and Wang, A. H.-J. (1984). *Annual Review of Biochemistry*, **53**, 791.
44. Wang, A. H.-J., Quigley, G. J., Kolpak, F. J., van der Marel, G., van Boom, J. H., and Rich, A. (1981). *Science*, **211**, 171.
45. Egli, M., Williams, L. D., Gao, Q., and Rich, A. (1991). *Biochemistry*, **30**, 11388.
46. Fujii, S., Wang, A. H.-J., Quigley, G. J., Westerink, H., van der Marel, G., van Boom, J. H., and Rich, A. (1985). *Biopolymers*, **24**, 243.
47. Wang, A. H.-J., Hakoshima, T., van der Marel, G., van Boom, J. H., and Rich, A. (1984). *Cell*, **37**, 321.
48. Wang, A. H.-J., Gessner, R. V., van der Marel, G. A., van Boom, J. H., and Rich, A. (1985). *Proceedings of the National Academy of Sciences, USA*, **82**, 3611.
49. Schneider, B., Ginell, S. L., Jones, R., Gaffney, B., and Berman, H. M. (1992). *Biochemistry*, **31**, 9622.
50. Rahmouni, A. R. and Wells, R. D. (1992). *Journal of Molecular Biology*, **223**, 131.
51. Tullius, T. D. and Dombroski, B. A. (1985). *Science*, **230**, 679.
52. Rhodes, D. and Klug, A. (1981). *Nature*, **292**, 378.
53. Marini, J. C., Levene, S. D., Crothers, D. M., and Englund, P. T. (1982). *Proceedings of the National Academy of Sciences, USA*, **79**, 7664.
54. Crothers, D. M., Haran, T. E., and Nadeau, J. G. (1990). *Journal of Biological Chemistry*, **265**, 7093.
55. Hagerman, P. J. (1990). *Annual. Review of Biochemistry*, **59**, 755.
56. Koo, H.-S., Drak, J., Rich, J. A., and Crothers, D. M. (1990). *Biochemistry*, **29**, 4227.

57. Nadeau,J.G. and Crothers,D.M. (1989). *Proceedings of the National Academy of Sciences, USA*, **86**, 2622.
58. Ulanovsky,L. and Trifonov,E.N. (1987). *Nature*, **326**, 720.
59. Goodsell,D.S., Kopka,M.L., Cascioo,D., and Dickerson,R.E. (1993). *Proceedings of the National Academy of Sciences, USA*, **90**, 2930.
60. Diekmann,S., Mazzaralli,J.M., McLaughlin,L.W., von Kitzing,E., and Travers,A.A. (1992). *Journal of Molecular Biology*, **225**, 729.
61. Arnott,S., Chandrasakaran,R., Hall,I.H., and Puigjaner,L.C. (1983). *Nucleic Acids Research*, **11**, 4141.
62. Chandrasakaran,R. and Radha,A. (1992). *Journal of Biomolecular Structure and Dynamics*, **10**, 153.
63. Aymami,J., Coll,M., Frederick,C.A., Wang,A.H.-J., and Rich, A. (1989). *Nucleic Acids Research*, **17**, 3229.
64. Fritsch,V. and Westhof,E. (1991). *Journal of the American Chemical Society*, **113**, 8271.

4

DNA–DNA recognition: non-standard DNA structures

1. Mismatches in DNA

1.1 General aspects

Base pairing in DNA is often considered solely in terms of Watson–Crick hydrogen bonding. Even though 16 distinct arrangements for A·T and G·C base pairs are in principle possible, few have actually been observed. The Hoogsteen and reverse Hoogsteen pairs (*Figure 4.1a* and *b*), have been found in several crystal structures of adenine:thymine and adenine:uracil complexes. These arrangements involve atoms N6 and N7 of adenine rather than the N1 and N6 atoms of Watson–Crick hydrogen bonding—so Hoogsteen hydrogen bonding with thymine is with a different edge of the adenine base, that which faces the major groove in B-DNA. Thus Hoogsteen pairing implies that the adenine base has to be in a *syn* glycosidic angle conformation if a Hoogsteen base pair is to be present in an anti-parallel DNA duplex.

Mismatched base pairing in DNA can arise naturally during replication. Mismatched pairs, i.e. G·A, G·T, A·A, G·G, T·T, C·C, T·C, or A·C, will, if left unchecked, lead to mutation and nonsense or incorrect gene products. The cell, however, has evolved a number of enzymatic repair mechanisms to recognize mispairs, whose efficiency depends on the nature of the mispairing. These

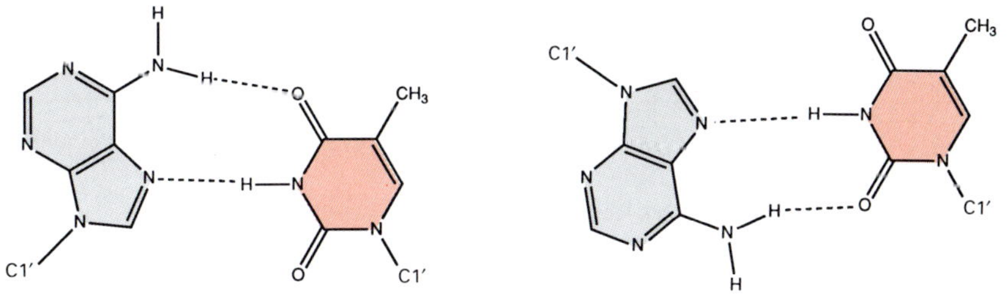

(a)   (b)

Figure 4.1. (a) Hoogsteen and (b) reverse Hoogsteen A·T base pair hydrogen bonding arrangements.

mechanisms have been extensively studied in *Escherichia coli*, but are much less well understood in eukaryotic organisms. In general, recognition is followed by duplex unwinding and excision of the mispair. Structural studies on DNA mispairs have addressed the questions of:

(1) the nature of the hydrogen bonding involved in the pairings;
(2) how these affect the local and global structure of duplex DNA;
(3) the relationship between structure, sequence context of the mispair and efficiency of mismatch repair.

Mismatches generally destabilize duplex DNA (this can be shown by the extent to which the temperature at which the transition from double helix to coil occurs, is decreased). Of the eight mispairs listed above, A·G has been the best studied and is the one that is focused upon here. Structural studies have been reported on oligonucleotides in A, B, and Z forms with G·T, I·T, U·G, I·C, and I·A base pairs (where I represents inosine).

Mutagenesis and/or carcinogenesis can occur when external agents such as methylating compounds produce covalent adducts with DNA bases. Until recently, little structural work has been reported on covalently modified oligonucleotides, in the past largely because of the difficulties involved in producing sufficient pure quantities of adducts for X-ray and NMR studies. These difficulties have now been largely overcome, and an increasing number of structural studies (especially using NMR techniques), have now been performed on oligonucleotides incorporating, for example, methylated bases (1). Of particular interest has been the mutagenic methylation of the O6 atom in guanine, which is produced by a variety of environmental and chemotherapeutic agents. Structural studies on these systems are described in Section 1.3.

1.2 Purine:purine mismatches

Crystal structures have been determined for a number of self-complementary oligonucleotide sequences with A·G base pairs (*Table 4.1*). These show a variety of base pairings (*Figure 4.2*). The arrangement with both nucleosides in an *anti*

Table 4.1   Purine:purine mismatched oligonucleotide crystal structures, with the mismatched base pairs shown in bold

Sequence	Glycosidic angles	C1′–C1′ distances (Å)	Reference
5′-CGCGAATTAGCG GCGATTAAGCGC	G(*anti*)·A(*syn*)	10.6, 10.8	2
5′-CGCAAATTGGCG GCGGTTAAACGC	A(*anti*)·G(*syn*)	10.8 (average)	3
5′-CGCAAGCTGGCG GCGGTCGAACGC	A(*syn*)·G(*anti*)	10.7 (average)	4
5′-CCAAGATTGG GGTTAGAACC	A(*anti*)·G(*anti*)	12.5	5
5′-CGCGAATTGGCG GCGGTTAAGCGC	G(*anti*)·G(*syn*)	10.7, 11.2	6

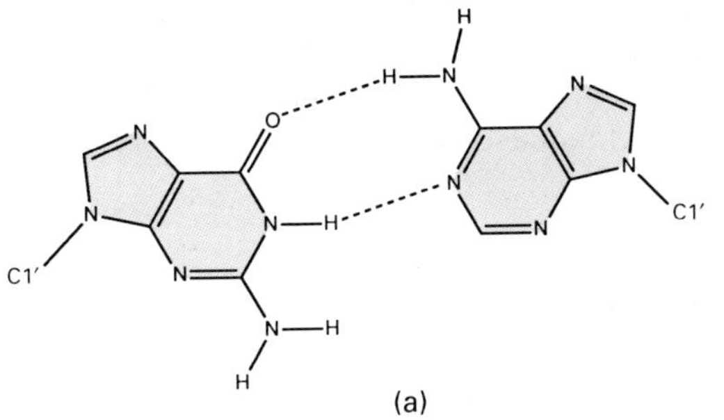

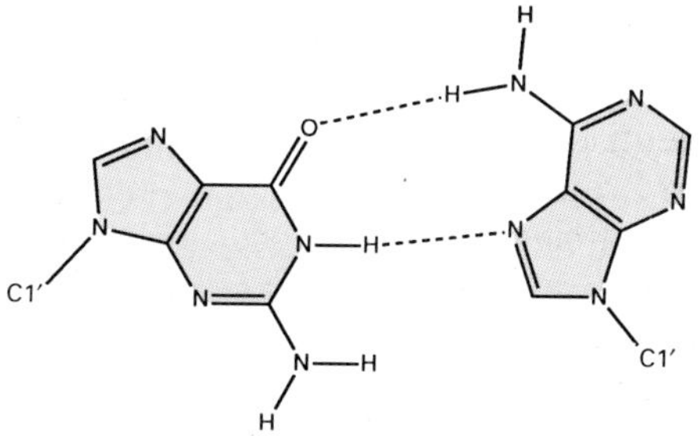

(a)

(b)

Figure 4.2.   (a) The G(*anti*)·A(*anti*) base pair. (b) The G(*anti*)·A(*syn*) base pair.

conformation necessarily forces the inter-strand C1'–C1' distance to be greater than in a standard B-DNA Watson–Crick base pair (10.4 Å). This *anti, anti* form (*Figure 4.2a*) has only been observed in one crystal structure (3), where there are two consecutive A·G base pairs. Presumably the bulge in the helix produced by their excessive C1'–C1' distances is relieved by the two A·G base pairs being well stacked on each other. For isolated purine:purine base pairs, it is easier to achieve normal C1'–C1' distances, and hence greater helix stability, by having one base *anti* and the other in a *syn* arrangement (*Figure 4.2b*). The base pairing is of the Hoogsteen type. Energetic considerations (Chapter 2) suggest that a *syn* conformation would be preferred for the guanine rather than the adenine base, but *Table 4.1* shows that this is only sometimes the case. All of the A·G-containing oligonucleotide structures have B-form structures, with generally standard geometries and few distortions from ideality, except that the mispairs tend to be poorly stacked with their neighbouring base pairs. Detailed examination of the stacking patterns has shown that the particular glycosidic conformation adopted by a mismatched base is such as to maximize the stacking, and so overrides the slight energy penalty of adenine being in a *syn* conformation. The nature of the sequence surrounding the mismatch is important, with most of the established structures following the rule that when the mispaired base on the first strand is at the centre of the three-residue sequence 5'-Py-**Pu**-Pu, then the central purine is in an *anti* conformation. An oligonucleotide with guanine:guanine base pairs has also been found (6) to have a B-like structure and an *anti, syn*

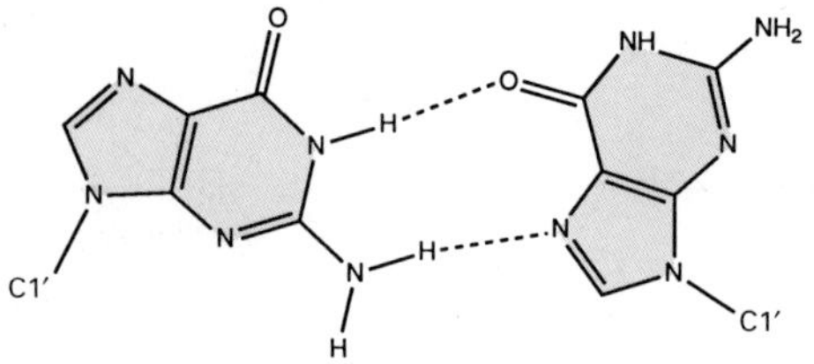

Figure 4.3. The G(*anti*)·G(*syn*) base pair.

arrangement for the mismatched bases (*Figure 4.3*). It appears that this unusual mismatch is associated with B_{II} backbone phosphate conformations (see Chapter 3, Section 3.1), which have also been documented for some adjacent G·A mismatches (7). The B_{II} conformation results in altered phosphate group intra-strand distances; these changes could act as recognition signals for repair nucleases, to identify and excise the mispairs at these points.

NMR studies of G·A-containing oligonucleotides have shown that the pattern of glycosidic angles seen in the crystal structures is not always maintained in solution, and some arrangements can be readily converted into others with changes in pH (8). Thus, the duplex of d(CGCAAATTGGCG)$_2$ adopts an A(*anti*)·G(*anti*) arrangement at high pH (9), yet a protonated A$^+$(*anti*)·G(*syn*) one at low pH. It should be borne in mind that the typical NMR experiment uses somewhat different conditions in the oligonucleotide solution compared with those in crystallizations, with higher salt concentrations in the former and high alcohol concentrations in the latter. These differences are likely to have an effect on conformational equilibria.

As has already been seen, the variability of G·A base pairs (and hence their stability) is largely dependent on the optimization of base stacking interactions, and hence on the nature of adjacent bases in a sequence. This is graphically illustrated in the sequence d(ATGAGCGAATA)$_2$, with four G·A base pairs (10). NMR and molecular modelling show that these have N7 . . . N2 and N3 . . . N2 hydrogen bonds, with only this arrangement being capable of stabilization by stacking interactions with adjacent bases. More generally, the unit 5'-Py-**GA**-Pu forms a particularly stable mispaired unit (11).

1.3 Alkylation mismatches

The methylation at the O6 position of guanines in DNA induces the resulting destabilizing G(OMe)·C base pair to undergo a transition mutation to a G(OMe)·T base pair. Methylation of the O6 carbonyl group makes the C6–O6 bond a single one, and hence alters the overall pattern of tautomerism in the guanine ring system, with N1 no longer having an attached hydrogen atom. NMR studies on the G(OMe)·C mispair in an oligomer duplex (12) indicate that both bases have *anti* glycosidic angles, in an overall B-type helix, with 'wobble' hydrogen bonding solely between N2(G) and O2(C), and with the O6 methyl group oriented towards the cytosine (*Figure 4.4a*). The crystal structure of a (Z-

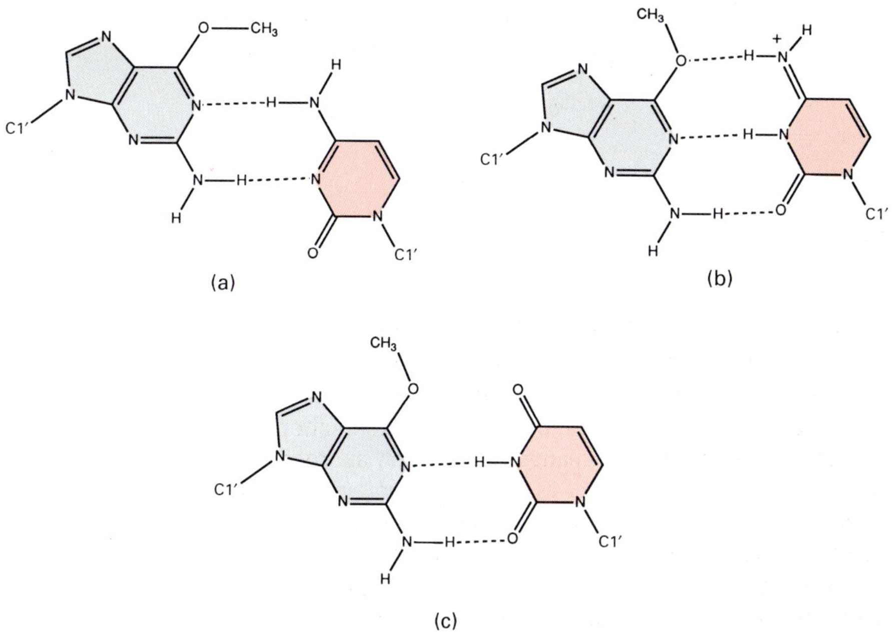

(a)   (b)

(c)

Figure 4.4   (a) The G(OMe)·C wobble base pair; (b) the G(OMe)·C Watson–Crick base pair; and (c) the G(OMe)·T base pair.

DNA) oligonucleotide duplex containing this mispair (13) shows an arrangement that is more compatible with standard Watson–Crick hydrogen bonding—distances O6(G). . .N4(C) and N1(G). . .N3(C) are normal hydrogen bonding ones (*Figure 4.4b*). This suggests that, even though the crystals were grown at pH 7.0, atoms N1(G) or N3(C) have acquired a proton so that partial Watson–Crick hydrogen bonding can occur, which is still less stable than a standard G·C base pair.

The G(OMe)·T base pair, which also destabilizes a duplex, shows Watson–Crick-type hydrogen bonding in a B-DNA dodecamer crystal structure (14), with probably two hydrogen bonds (*Figure 4.4c*). The overall shape of this base pair is close to that of a standard G·C one, whereas the G(OMe)·C one is rather less so, especially in the wobble arrangement (*Figure 4.4a*). This difference may explain the observed preference for G(OMe)·T base pairs to be incorporated into DNA during replication and not to be readily recognized by repair enzymes.

2. Multi-stranded DNA helices

2.1 The triple helix
The formation of a triple helix (*Figure 4.5*) by two pyrimidine strands and one purine strand was discovered (15) soon after the structure of the double helix

Figure 4.5. Schematic view of a nucleic acid triple helix, with the third strand shaded orange and the first two duplex strands shown shaded grey.

itself was determined, when solutions of poly(A) and poly(U) were mixed in appropriate proportions, forming a 1:1:1 three-stranded polynucleotide complex, poly(A·U·U). A molecular model for this novel helix was proposed on the basis of fibre diffraction data (16) for both deoxy- and ribo-polynucleotides. The triple helix is right-handed, with an (adenine) purine strand Watson–Crick hydrogen bonding to a (thymine or uracil) pyrimidine one, and a third (thymine or uracil) pyrimidine strand Hoogsteen hydrogen bonding to the purine strand and parallel to it (*Figures 4.6a* and *4.7*). This hydrogen bonding arrangement is termed a base triplet and the triple helix itself is often termed a triplex. The third strand in the triple helix occupies the major groove of the purine–pyrimidine double helix, which has some (though not all) features of classic A-DNA; it has bases significantly (~3.5 Å) displaced from the helix axis, it has an average helical twist of ~30°, and the duplex minor-groove width, 10.7 Å, is almost identical to the A-DNA one of 11.0 Å. However, its large major-groove width of 9.8 Å, which is necessary to accommodate the third strand, is much greater than in A-DNA. Triple helices can be formed by relatively short oligonucleotides in solution, such as $d(A)_{12} \cdot 2d(T)_{12}$, and a number of NMR studies have been performed (17–19) on such systems as well as intramolecular triplexes formed by a single strand of

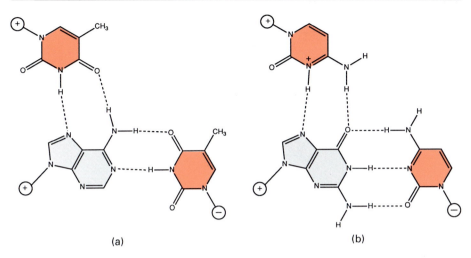

Figure 4.6.   (a) The T·A·T hydrogen bonding arrangement in a Py·Pu·Py triple helix, with strand directions indicated by signs. (b) The C$^+$·G·C arrangement in the Pu·Py·Pu triple helix.

appropriate sequence folding back on itself. There is as yet no single-crystal X-ray structure of a triple helix, but several molecular mechanics and dynamics studies have been performed (20, 21) which, together with the NMR studies, confirm in outline the features of the fibre-diffraction model.

One guanine-containing and two cytosine-containing strands can form a triple helix, which has the same polarity of strands as the T·A·T triplex. However, now third-strand cytosine is required to be protonated at the N3 position in order for effective Hoogsteen hydrogen bonding to take place (*Figure 4.6b*). Such protonation can only take place at about pH 5.0, placing a severe practical limitation on C·G·C triplex formation in biological systems (see below). Substitution of a methyl group or bromine at the 5-position of cytosine results in triplex formation being possible for C·G·C-containing sequences at approximately pH 7 (i.e. around physiological pH). The precise reasons for this are not clear; it has been suggested that there is an increased hydrophobic contribution to third-strand binding when there are these substituents at the 5-position of cytosine, rather than there being a shift in the pK_a of the N3 atom of cytosine. Triple helices can be formed with both cytosines and thymines in the third strand, that is with mixed pyrimidine sequences. Such a sequence (5′-TTCTTTTCTTCTTTCTTTTT), with a co-valently attached strand cleavage agent, has been used to bind to and cleave a unique site on a yeast chromosome (22), demonstrating the exceptionally high specificity for a target duplex sequence that triplex formation can achieve.

Triple helices can also be formed with the third strand having solely purines, so that the resulting helix has Pu·Pu·Py triplets of bases (23). Such a triple helix forms most readily with G·G·C base triplets, having a G·G hydrogen bonding arrangement that is likely to be as shown in *Figure 4.8a* and *b*, with the third,

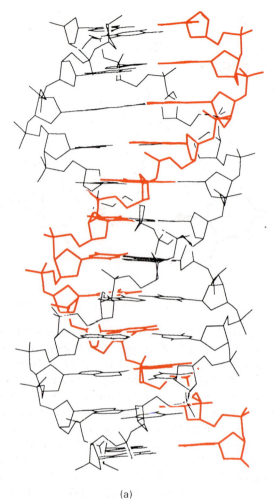

(a)

Figure 4.7.   Two views of a molecular structure for the Py·Pu·Py triple helix. In (a) the third (pyrimidine) strand is shown with orange bonds. (b) A more detailed view, now of two base-triplets in the structure, looking down on to the planes of the bases. Hydrogen bonds are shown with dotted lines.

purine strand being anti-parallel to the purine strand of the duplex. This arrangement is pH-independent, contrasting with the C·G·C triple helix. The guanosine nucleosides on the third strand can have either *anti* or *syn* arrangements (*Figure 4.8a* or *b*); both NMR (24) and molecular modelling studies (25) suggest that the *anti* one is in fact preferred.

There have been numerous attempts to extend the sequence restrictions on triple-strand recognition. For example, a thymine in a pyrimidine third strand can be replaced by a guanine, with the resultant G·T·A base triplet being relatively stable (26), although the distortions that it is likely to produce in the DNA

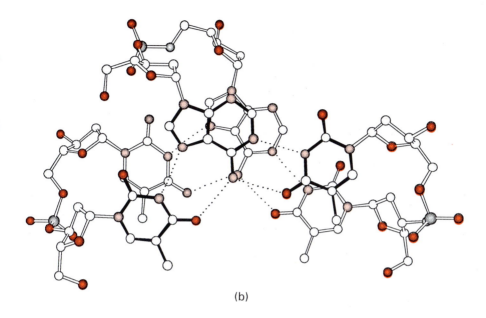

(b)

backbone suggest a limit on the number of such triplets that can be tolerated within a sequence. Almost all other possible triplet mismatches are highly destabilizing to triple helix formation (27, 28) such that triplex formation for an otherwise stable sequence can be abolished by even a single such mismatch. The nature of the flanking sequence is important for a few mismatches, such as T·C·G, that may be stable in some circumstances. More general recognition of all four Watson–Crick base pairs in a duplex (i.e. C·G, G·C, A·T, and T·A) is almost certainly only achievable with non-standard 'designer' bases (29) that are capable of being incorporated into a general-sequence oligonucleotide and hybrid-izing with a target duplex with high affinity. Yet more radical are the 'peptide' nucleic acids in which the conventional sugar–phosphate backbone has been totally replaced by repeating peptide groups of precise structural equivalence (30) yet bind much more avidly to duplexes than do conventional third strands.

Much of the interest in the triplex phenomenon arises from the findings that triplex formation can specifically inhibit transcription of a particular gene (31), by binding to a promoter region that contains a suitable triplex-forming sequence. The (unsurprising) rarity of suitable sequences has encouraged studies aimed at extend-ing the sequence restrictions on triplex formation that have been outlined above, so as to be able to recognize a regulatory sequence for any gene that is a suitable therapeutic target. This 'antigene' approach has been found to work in cells, and so has much promise as a way of switching off, for example, particular oncogenes.

2.2 Guanine quadruplexes

It has long been known that runs of guanosine nucleosides in oligo- and poly-nucleotides can readily aggregate together, provided a monovalent cation such as

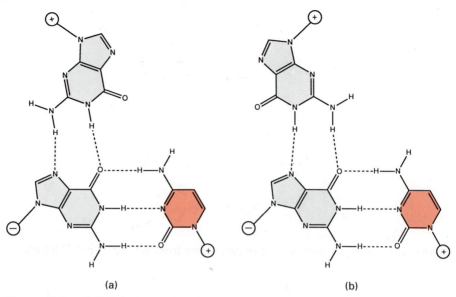

Figure 4.8.   G·G·C hydrogen bonding arrangements in a Pu·Py·Pu triple (a) has the third-strand G in an *anti* conformation, in (b) it is *syn*.

potassium is present to provide stabilization. Diffraction patterns from fibres of poly(G) have been interpreted as arising from a novel four-stranded helix (32), with four guanine bases all involved in a tetrad G·G base pairing arrangement. This could involve Hoogsteen hydrogen bonding analogous to that observed in G·G mismatched duplexes, with therefore an alternating pattern of *syn* and *anti* nucleosides within any one level of tetrad and anti-parallel strands (*Figure 4.9*), although parallel, all-*anti* arrangements are also possible in principle.

Guanine-rich sequences occur as overhangs on the 3' ends of eukaryotic chromosomes (known as telomeres), with repeats of guanines and adenines/thymines, typically of the type G_nT_n, $G_nT_nG_n$, or G_nA_n. Simple sequences of this type can dimerize to form guanine tetrads (33) between hairpin loops that contain unpaired thymines and are formed at the 3' ends of the sequence (*Figure 4.9*), and are therefore relevant models for telomeres. It appears that in principle a number of isomers of such a structure can exist, which differ in the orientations of glycosidic angles within the various strands of the complex. Whether strands are parallel or anti-parallel might depend in part on the nature of the sequence, or even the counter-ion used.

A crystallographic analysis of the sequence d(GGGGTTTTGGGG) has shown that this sequence forms a four-stranded, folded quadruplex (34) from two molecules associating together, with hydrogen bonding of the guanines as shown in *Figure 4.9*. Each sequence has formed an intramolecular hairpin structure—the overall complex then has all four strands anti-parallel to each other (*Figure 4.10a*), and an alternation of *syn* and *anti* glycosidic angles along a chain. The four

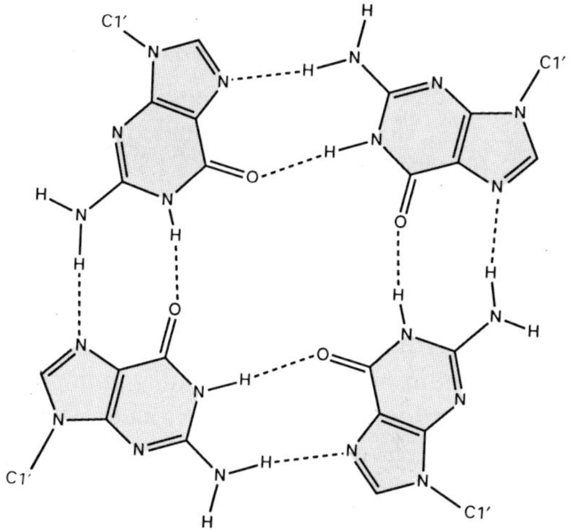

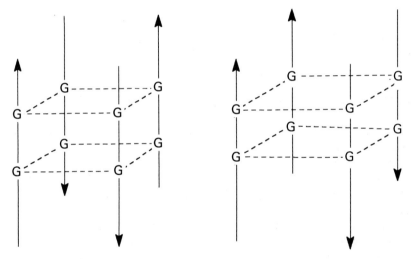

Figure 4.9.   The arrangement of guanine:guanine hydrogen bonds in a G-tetrad.

Figure 4.10.   Possible (a) anti-parallel and (b) parallel arrangements for four G_n strands in G-tetrad structures.

tetrad-guanine groups ('G-quartets') are sandwiched on top of each other at a 3.4 Å separation. It is intriguing that an NMR study (35) on this sequence did not reveal an anti-parallel arrangement for all four strands. Instead, the adjacent strands are alternately parallel and anti-parallel (*Figure 4.10b*). There are still alternating *syn* and *anti* glycosidic conformations along any one strand, but the arrangement of strands results in a G-quartet arrangement with a *syn-syn-anti-anti* pattern of glycosidic angles. The hydrogen bonding arrangement in the quartet itself remains as in the X-ray structure. The fact that quite distinct topological arrangements are observed for a given tetraplex structure suggests that at least some telomere sequences can be satisfactorily folded in more than one way, probably depending on the precise environmental conditions. Other sequences have been studied by NMR methods; some, such as d(TGGGGT) (36) and d(TTGGGG) (37), form parallel-stranded quadruplex structures with *anti* glycosidic angles. Others, such as d(GGTTTTCGG) (38), form anti-parallel arrangements with alternating *syn-anti* glycosidic bonds.

Some general patterns of the relationships between strand polarity and base pairing, have emerged from studies on triplexes and quadruplexes, as well as on Watson–Crick, left-handed, and mispaired duplexes (39, 40). For example, the pattern of glycosidic angles for a base pair is a consequence of the relationship between strand directions in quadruplexes, with *syn* angles being produced by anti-parallel strands and *anti* angles by parallel ones. This has led to a general classification scheme for nucleic acid structural types (40) based on these polarity and base pairing criteria. Prediction of new structural types emerges naturally from the scheme; the variety of DNA structures that has been established over the past few years does suggest that at least some of these predictions will actually exist.

3. Further reading

Mispairing:

Brown,T. and Kennard,O. (1992). *Current Opinion in Structural Biology*, **2**, 354.
Hunter,W.N. (1992). *Methods in Enzymology*, **211**, 221.
Modrich,P. (1987). *Annual Review of Biochemistry*, **56**, 435.
Modrich,P. (1991). *Annual Review of Genetics*, **25**, 229.
Singer,B. and Grunberger,D. (1983). *Molecular biology of mutagens and carcinogens*. Plenum Press, New York.

Triplexes:

Hélène,C. and Toulmé,J.J. (1990). *Biochimica et Biophysica Acta*, **1040**, 99.
Moser,H. and Dervan,P.B. (1987). *Science*, **238**, 645.
Wells,R.D., Collier,D.A., Hanvey,J.C., Shimizu,M., and Wohlraub,F. (1988). *FASEB Journal*, **2**, 2939.

Quadruplexes:

Blackburn,E.H. (1990). *Journal of Biological Chemistry*, **265**, 5919.
Blackburn,E.H. (1991). *Nature*, **350**, 569.

4. References

1.   Patel,D.J. (1992). *Current Opinion in Structural Biology*, **2**, 345.
2.   Brown,T., Hunter,W.N., Kneale,G., and Kennard,O. (1986). *Proceedings of the National Academy of Sciences, USA*, **83**, 2402.
3.   Leonard,G.A., Booth,E.D., and Brown,T. (1990). *Nucleic Acids Research*, **18**, 5617.

4. Webster, G.D., Sanderson, M.R., Skelly, J.V., Neidle, S., Swann, P.R., Li, B.F., and Tickle, I.J. (1990). *Proceedings of the National Academy of Sciences, USA*, **87**, 6693.
5. Privé, G.G., Heinemann, U., Chandrasegaran, S., Kan, L.-S., Kopka, M.L., and Dickerson, R.E. (1987). *Science*, **238**, 498.
6. Skelly, J.V., Edwards, K.J., Jenkins, T.C., and Neidle, S. (1993). *Proceedings of the National Academy of Sciences, USA*, **89**, 804.
7. Chou, S.-H., Cheng, J.-W., and Reid, B.R. (1992). *Journal of Molecular Biology*, **228**, 138.
8. Gao, X. and Patel, D.J. (1988). *Journal of the American Chemistry Society*, **110**, 5178.
9. Lane, A., Jenkins, T.C., Brown, D.J.S., and Brown, T. (1991). *Biochemical Journal*, **279**, 269.
10. Li, Y., Zon, G., and Wilson, W.D. (1991). *Proceedings of the National Academy of Sciences, USA*, **88**, 26.
11. Li, Y., Zon, G., and Wilson, W.D. (1991). *Biochemistry*, **30**, 7566.
12. Kalnik, M.W., Li, B.F.L., Swann, P.F., and Patel, D.J. (1989). *Biochemistry*, **28**, 6182.
13. Ginell, S.L., Kuzmich, S., Jones, R.A., and Berman, H.M. (1990). *Biochemistry*, **29**, 10461.
14. Leonard, G.A., Thomson, J., Watson, W.P., and Brown, T. (1990). *Proceedings of the National Academy of Sciences, USA*, **87**, 9573.
15. Felsenfeld, G., Davies, D.R., and Rich, A. (1957). *Journal of the American Chemical Society*, **79**, 2023.
16. Arnott, S., Bond, P.J., Selsing, E., and Smith, P.J.C. (1976). *Nucleic Acids Research*, **11**, 4141.
17. de los Santos, C., Rosen, M., and Patel, D.J. (1989). *Biochemistry*, **28**, 7282.
18. Sklenár, V. and Feigon, J. (1990). *Nature*, **345**, 836.
19. Pilch, D.S., Levenson, C., and Shafer, R.H. (1990). *Proceedings of the National Academy of Sciences, USA*, **87**, 1942.
20. Cheng, Y.-K. and Pettitt, B.M. (1992). *Journal of the American Chemical Society*, **114**, 4465.
21. Laughton, C.A. and Neidle, S. (1992). *Journal of Molecular Biology*, **223**, 519.
22. Strobel, S.A. and Dervan, P.B. (1990). *Science*, **249**, 73.
23. Beal, P.A. and Dervan, P.B. (1991). *Science*, **251**, 1360.
24. Radhakrishnan, I., de los Santos, C., and Patel, D.J. (1991). *Journal of Molecular Biology*, **221**, 1403.
25. Laughton, C.A. and Neidle, S. (1992). *Nucleic Acids Research*, **20**, 6535.
26. Griffin, L.C. and Dervan, P.B. (1989). *Science*, **245**, 967.
27. Mergny, J.-L., Sun, J.-S., Rougée, M., Montenay-Garestier, T., Barcelo, F., Chomilier, J., and Hélène, C. (1991). *Biochemistry*, **30**, 9791.
28. Kiessling, L.L., Griffin, L.C., and Dervan, P.B. (1992). *Biochemistry*, **31**, 2829.
29. Griffin, L.C., Kiessling, L.L., Beal, P.A., Gillespie, P., and Dervan, P.B. (1992). *Journal of the American Chemical Society*, **114**, 7976.
30. Nielsen, P.E., Egholm, M., Berg, R.H., and Buchardt, O. (1991). *Science*, **254**, 1497.
31. Grigoriev, M., Praseuth, D., Guieysse, A.L., Robin, P., Thoung, N.T., Hélène, C., and Harel-Bellan, A. (1993). *Proceedings of the National Academy of Sciences, USA*, **90**, 3501.
32. Arnott, S., Chandrasekaran, R., and Marttila, C.M. (1974). *Biochemical Journal*, **141**, 537.
33. Sundquist, W.I. and Klug, A. (1989). *Nature*, **342**, 825.
34. Kang, C., Zhang, X., Ratliff, R., Moyzis, R., and Rich, A. (1992). *Nature*, **356**, 126.
35. Smith, F.W. and Feigon, J. (1992). *Nature*, **356**, 164.

36. Aboul-ela,F., Murchie,A.I.H., and Lilley,D.M.J. (1992). *Nature*, **360**, 280.
37. Wang,Y. and Patel,D.J. (1992). *Biochemistry*, **31**, 8112.
38. Wang,Y., de los Santos,C., Gao,X., Greene,K., Live,D., and Patel,D.J. (1991). *Journal of Molecular Biology*, **222**, 819.
39. Westhof,E. (1992). *Nature*, **358**, 459.
40. Lavery,R., Zakrzewska,K., Sun,J.-S., and Harvey,S.C. (1992). *Nucleic Acids Research*, **20**, 5011.

5

Principles of ligand–DNA recognition

1 Introduction

The recognition of a DNA molecule can, on the one hand, be general for any individual nucleotide or sequence. At the other extreme, it can be highly specific, solely recognizing a defined sequence. This can be done very effectively by DNA itself, as has been seen in previous chapters. A single strand of oligonucleotide can recognize another strand, a duplex (to give a triplex) or even three other strands (forming a quadruplex), with in each case an increasingly stringent sequence requirement. The previous chapter has described the features of DNA–DNA recognition, and the fact that it is governed by base–base hydrogen bonding. This is the manner whereby direct readout of DNA sequence information can be achieved, by small molecules and proteins as well as by DNA itself. In order for direct readout to occur, an incoming molecule has to have access to the base pairs via a DNA groove. The orientation of this incoming group is important since hydrogen bonding, and hence direct readout, are directional interactions. The edges of the bases have hydrogen bonding potential (*Figure 5.1*) that is only partially satisfied by base pairing itself (1), and therefore is available for recognition by ligands with hydrogen bonding complementarity. This pattern of hydrogen bond donation and acceptance is sequence-dependent. Thus, in the major groove, a C·G Watson–Crick base pair has a pattern (donor, acceptor, acceptor) that is distinct from that for the reversed G·C base pair. The A·T and T·A base pairs have identical major groove donor/acceptor patterns, but the presence of the methyl group of thymine introduces an asymmetry into the groove that can enable effective discrimination between these two sequences of base pairs. Patterns of hydrogen bonding in the minor grooves of each pair of sequences are symmetric, so discrimination on this basis alone is not possible. The narrowness of the minor groove in B-type DNA sequences makes it unlikely that all three hydrogen bonds in a C·G or G·C base pair are simultaneously available, for purely steric reasons. More generally, accessibility of the bases, in whatever grooves are formed by the structure, is necessary for direct hydrogen bonding readout. This accessibility is a consequence of groove widths, which are them-

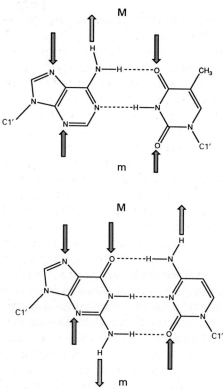

Figure 5.1. The pattern of hydrogen bond donors and acceptors in A·T and G·C hydrogen bonds, with the arrows indicating the directionality of the hydrogen bonds; M and m indicate major and minor grooves, respectively.

selves sequence-dependent in both static and dynamic ways, as discussed in Chapter 3. It has been established from the crystal structures of oligonucleotides that AT-rich minor grooves tend to be narrower than average; their widths, depend both on the length and the nature of the AT sequence. Other sequence-dependent features of base pairs and base pair steps, which can be both localized (such as propeller and helical twist) and long-range (typified by the bending of A tracts), will affect the geometric relationships between hydrogen bond donors/ acceptors on successive bases.

Indirect sequence readout by definition involves interaction with elements of DNA structure other than the base pair hydrogen bonds, and is consequently inherently less directional. The bases themselves provide several mechanisms for such readout. Base pair and step sequence-dependent structural features have been outlined in Chapter 3. There are also significant differences in electronic character between A·T and G·C base pairs, with the latter being more electron-rich. Thus, electron-deficient planar molecules that can stack with individual base pairs (such as intercalating drug molecules or tryptophan side-chains), will tend to bind preferentially to G·C base pairs. Such preferences are by their nature localized to individual base pair sites. However, the tendency for there to be less base stacking in B-DNA between pyrimidine-3′,5′-purine dinucleoside steps than purine-3′,5′-pyrimidine ones, means that the dinucleoside sequences CpG, TpG, CpA, and TpA (of the pyrimidine-3′,5′-purine type) are more readily unstacked and therefore planar molecules bind intercalatively to them in preference. The methyl group of thymine can play an important role in sequence readout, by being able to make stable van der Waals non-bonded contacts with hydrophobic side-chains such as those of leucine, isoleucine, and alanine. Indirect readout can also occur when the presence of a particular group, substituent, or side-chain would result in the occurrence of steric clashes, and therefore destabilization of the resulting DNA–ligand complex. An example would be the bulky methyl group of thymine, which would be allowable only in some circumstances.

Indirect readout of sequence-dependent structure can also take advantage of differences in backbone and phosphate conformations such as B_{II} phosphate geometries (Chapter 3), which can act as recognition signals. Perhaps more important are differences in distances between phosphate groups that necessarily result from backbone conformational differences, which can be recognized by hydrogen bonding/electrostatic interactions with basic amino acid side-chains.

The electronic characteristics of an individual base pair produce an electric field and potential in its vicinity. These electronic features are distinct for the various classical DNA structural types, as well as for changes in oligonucleotide sequences (2). Of especial significance is the finding that for B-DNA an electrostatic potential minimum occurs in the major groove of oligo(dG)·(dC) sequences, the minor groove of oligo(dA)·(dT) sequences and the 5′-AATT sequence in the Dickerson–Drew dodecamer. These all have more negative potentials than those around the phosphate groups, which explains why cationic ligands and basic protein side-chains cluster in the grooves rather than around the phosphates. This electrostatic factor is important in directing an incoming ligand to a region of sequence; small local variations in net charge and field around an individual base pair then could assume some importance, especially for electrophilically driven covalent bonding.

2. DNA–water interactions

The hydration of DNA is essential for the maintenance of its structural integrity. At least 9 or 10 first-shell water molecules are associated with each nucleotide. It has been presumed that some of these are clustered around phosphate groups, together with counter-ions such as sodium, potassium, or ammonium. One would expect other water molecules to be associated with the bases, possibly taking advantage of the hydrogen bonding offered by the edges of the bases (1). Fibre diffraction analyses have achieved some success in locating the positions of some counter-ions, but only with the advent of single-crystal analyses of oligonucleotides has it been possible to define the positions of water molecules. NMR methods are beginning to provide valuable complementary information, by locating some of the least mobile water molecules, such as the minor groove spine of hydration in the dodecamer d(CGCGAATTCGCG)$_2$ (3, 4). In general, the number of water molecules located by crystallography is highly dependent on the resolution of the structure and the thermal motion of individual water molecules (as well as on the quality of the raw X-ray diffraction data), although even high resolution ($\geq$1.5 Å) analyses cannot find more than a fraction beyond the first hydration shell. The nature and extent of DNA hydration also provide an indication of accessibility to ligands other than water itself, which are borne out by calculations of the solvent-accessible surface characteristics of DNA. These have shown (5) that there are some differences between A and B helical types in terms of available surface area in the grooves, as well as between A·T and G·C base pairs. Sequence-dependent backbone conformational variability will also alter accessibility—the narrow minor groove in A·T regions is sufficient for only a width of one water molecule per base pair. The corresponding network of water molecules (the spine of hydration) is necessarily then one-dimensional in form. The B-DNA major groove is dominated by the large, hydrophobic thymine methyl group—it is thus unsurprising that water molecules have not been observed to form structured networks in this groove. A systematic examination of the observed distribution of water molecules in known oligonucleotide crystal structures (6) has shown that the hydration of individual bases depends markedly on the helical type (A, B, or Z), as well as the nature of the groove.

The hydration patterns around phosphate groups in oligonucleotide crystal structures are dependent on whether the helix is an A- or B-type (7). In the former, successive phosphate groups along the phosphodiester backbone are close enough together for water molecules to be able to form bridges between them, as indeed is observed. Such bridging patterns are generally not found in B-form oligonucleotides, with their more separated phosphate groups. These water molecules prefer to hydrogen bond to the anionic phosphate oxygen atoms, whereas the ester oxygen atoms O3' and O5' are rarely hydrated. By contrast the deoxysugar ring oxygen atom frequently participates in water networks, especially in A-form structures. Several A-DNA octamer high-resolution structures have extended major groove networks of pentagonal-linked water molecules (8) whose co-ordination to bases appears to be a major factor in maintaining the integrity of the oligonucleotide structure.

3. DNA–drug recognition

A number of clinically important drugs and antibiotics are believed to exert their primary biological action by means of DNA interaction and subsequent inhibition of template function, such as the selective inhibition of DNA-directed RNA synthesis by actinomycin D. For many anti-tumour, intercalating drugs, drug–DNA interaction *in vivo* involves concomitant interactions with the enzyme DNA topoisomerase II to form a ternary complex. This eventually leads to lethal DNA double-strand breaks. The desire to understand the molecular basis for the action of such drugs, and rationally to design new ones, has been the impetus for many of the large number of structural studies in this field.

DNA-interactive drugs can be conveniently categorized into two major classes according to their mode of interaction. There are sub-divisions within each, which frequently overlap.

3.1 Non-covalent binding

Electrostatic-dominated binding involves sequence-neutral interactions between a cationic group and negatively charged phosphates on DNA. Examples of such ligands include the natural polyamines such as spermine and spermidine. The majority of non-covalent binding drugs carry formal positive charge, so electrostatic contributions are always a significant component of drug binding.

3.1.1 Intercalative binding

There are a large number of antibiotics, antibacterial and anti-tumour drugs that are characterized by the possession of an extended electron-deficient planar aromatic ring system (chromophore). This essential structural feature is typically two or three six-membered rings in size (*Figure 5.2*), approximately the same size as a base pair itself. The intercalation hypothesis, originally proposed by Lerman (9) suggests that the planar chromophore of the drug molecule becomes inserted in between adjacent base pairs in an intercalative manner (*Figure 5.3*). The drug chromophore is stabilized by van der Waals dispersion interactions between its planar group and the base pairs surrounding it. This results in an extension of the double helix by 3.4 Å per bound drug molecule, together with changes in helical twist (unwinding) for the base pairs at and adjacent to each binding site (10). The base pairs either side of an intercalated drug thus become 6.8 Å apart. DNA–intercalator recognition itself is essentially sequence-neutral, although, as described in Section 1, stacking requirements usually produce a small preference for pyrimidine-3′,5′-purine sequences. 'Simple' intercalators consist solely of an intercalating group, often carrying a positive charge—for example, proflavine and ethidium bromide.

The degree of helix unwinding produced by intercalation is dependent on the nature of the bound drug. It can be measured experimentally (10) using covalently-closed circular DNA, with reference to the standard value of 26° per bound drug molecule, for ethidium bromide. Crystallographic analyses of intercalation complexes have shown qualitative unwinding; however, the helical twist angle for the two base pairs immediately surrounding an intercalated drug is not necessarily equivalent to this angle. The ribodinucleoside $(CpG)_2$ duplex complex with proflavine (11) and the hexamer duplex $(CGTACG)_2$ with bound daunomycin (12) both have normal B-DNA helical twist angles of 36° for the base pairs flanking the drug. In the latter case, there are changes in helical twist at adjacent base pair steps. The unwinding angle is thus the sum of cumulative changes in helical twist over all affected base pairs. The process of base pair separation to produce an

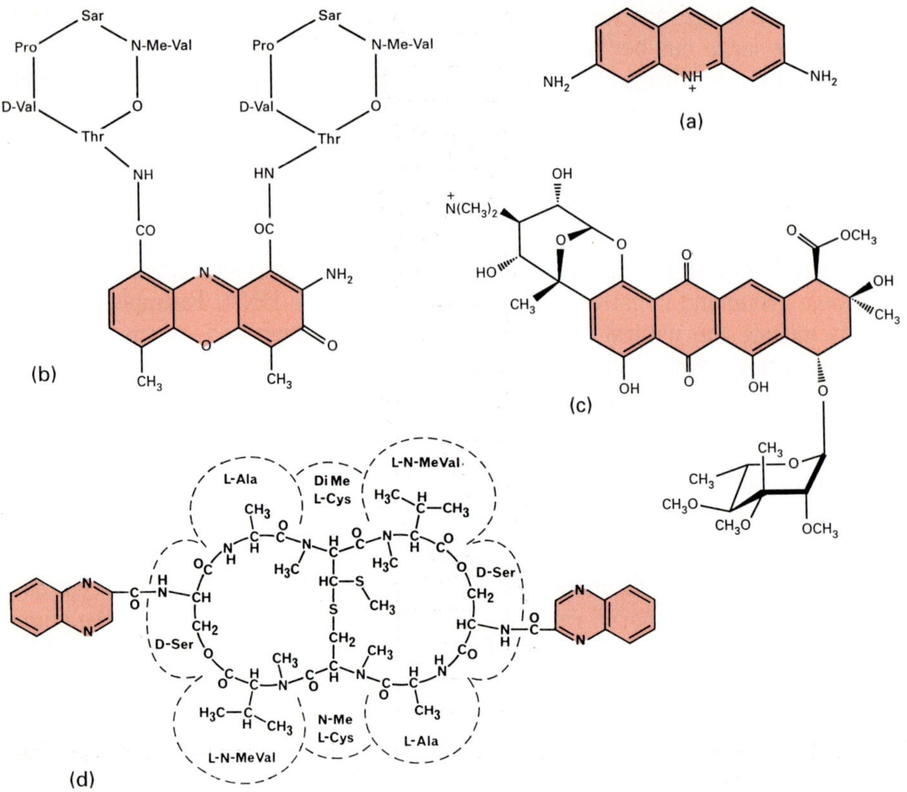

Figure 5.2. Structures of some intercalating molecules (a) proflavine, (b) actinomycin, (c) nogalamycin, and (d) echinomycin.

intercalation site also results in a number of changes in backbone conformation and base pair and base step geometry.

More complex intercalators than ethidium or proflavine—such as the anti-tumour drugs actinomycin and daunomycin—have attached groups such as side-chains, sugar rings, or peptide units. These groups reside in a DNA groove, where they are stabilized by van der Waals interactions, and can hydrogen bond to adjacent bases, providing sequence-specific direct readout. For example, the threonine residue in the cyclic pentapeptide part of the actinomycin molecule (*Figure 5.2*) hydrogen bonds to the N2 and N3 atoms of a guanosine nucleoside, which is then structurally constrained to be on the 5′-side of the intercalating drug chromophore, giving a requirement for the sequence GpX. This has been shown by an X-ray analysis of actinomycin bound to the sequence d(GAAGCTTC)$_2$, which finds the drug phenoxazone chromophore intercalated at the GpC site (13) and the pentapeptides situated in the minor groove of the DNA helix, in agreement with an NMR study with the sequence d(AAAGCTTT)$_2$ (14).

NMR and crystallographic studies on oligonucleotide complexes of the anti-

Figure 5.3.   Schematic representation of two drug molecules (shaded) intercalated into a DNA double helix.

tumour antibiotic nogalamycin (*Figure 5.2.*) have revealed a novel binding mode for this drug, with the two attached groups residing one in each groove; the drug chromophore itself is intercalated as expected (15, 16). The aglycone group is held in the minor groove and the amino sugar in the major groove (*Figure 5.4*), with drug–oligonucleotide hydrogen bonding to the N2 and O6 atoms of guanines, thereby providing simultaneous major and minor groove direct sequence recognition, which concurs with DNA footprinting data (17). In contrast with the simpler intercalators, the mechanism of nogalamycin intercalation into duplex DNA is not straightforward. The two attached groups, one at each end of the molecule, are sufficiently bulky to ensure that the drug chromophore cannot become intercalated without one or other of these groups passing through the intercalation site. This implies that drug–DNA association and dissociation will both be exceptionally slow processes as a direct result of the structural constraints forced on the complex.

Yet more complex are the drug molecules typified by the echinomycin family of anti-tumour antibiotics (*Figure 5.2*), with two chromophores linked together by cyclic oligopeptides or other groupings. These molecules bis-intercalate, with the chromophores, at a separation of 10.2 Å, simultaneously binding at sites

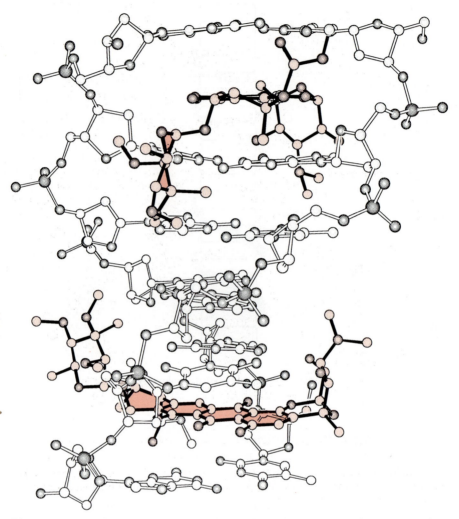

Figure 5.4. A view of the structure of a nogalamycin–hexanucleotide complex, from a single-crystal analysis (ref. 15). The drug molecules are shown with solid bonds.

separated by two intervening base pairs (*Figure 5.5*), as shown by crystallographic and NMR studies (18–21) on echinomycin and the closely related compound triostin A bound to oligonucleotide sequences. Echinomycin itself has a sequence requirement for 5′-CpG, which is flanked by the two quinoxaline chromophores. The peptide ring system sits in the DNA minor groove, with specific hydrogen bonding to the 2-amino group of the guanine (22). An unexpected finding in the crystal structures of the drug complexes has been the occurrence of Hoogsteen hydrogen bonding for the base pairs flanking the bound drug—the two base pairs between the chromophores are always in a standard

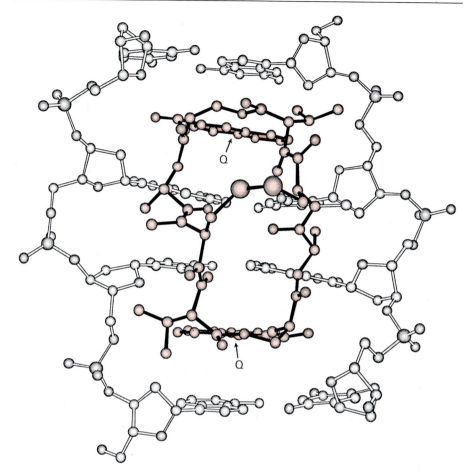

Figure 5.5. A view of part of the crystal structure of the drug triostin A bis-intercalated between two G·C base pairs (ref. 19). The quinoxaline chromophores of the drug are marked 'Q'.

Watson–Crick arrangement. NMR studies indicate that this major structural change from standard B-DNA can occur with these drugs and certain short oligonucleotide sequences in solution. However chemical probe experiments with much longer (160 bp) sequences that are more truly representative of biological DNA, strongly suggests that Hoogsteen base pairing does not occur in these, and thus does not need to accompany the structural changes consequent to bis-intercalative recognition.

3.1.2 Groove binding

A number of compounds (*Figure 5.6*) that generally show a preference for AT regions can be classified as groove-binders (23, 24). By contrast with intercalat-

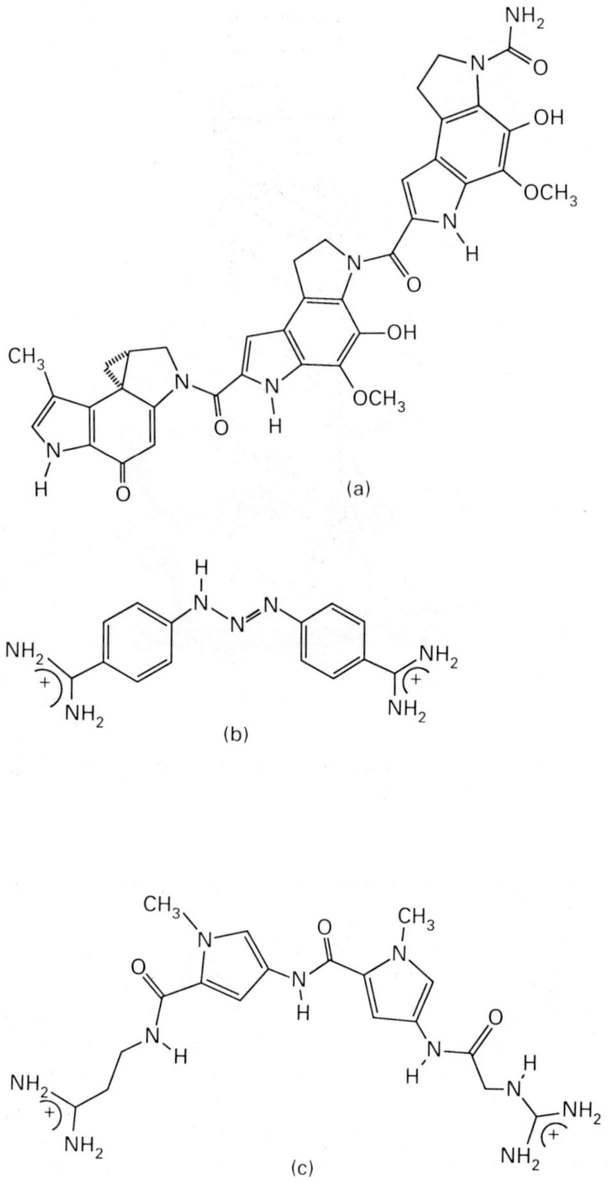

Figure 5.6. Structures of some groove-binding drugs (a) CC1065, (b) berenil, (c) netropsin.

ing drugs, they do not significantly perturb DNA structure (*Figure 5.7*). They bind exclusively in the minor groove of B-DNA duplexes. The mitochondrial DNA of a number of organisms has extended regions of AT sequence and it has been suggested that these drugs preferentially bind in such regions, thereby

Figure 5.7.   Schematic representation of two drug molecules (shaded) bound in the minor groove of a DNA double helix.

selectively inhibiting their respiratory functions. Several of them find medicinal use as anti-parasitic or anti-viral agents.

Crystallographic and NMR studies on complexes of these drugs with, for example, the dodecamer duplex d(CGCGAATTCGCG)$_2$ and closely-related ones, show that drugs such as berenil (25, 26) and netropsin (27, 28) are situated in the narrow minor groove AT regions of these sequences (*Figure 5.8*). The drug molecules are stabilized in position by close van der Waals interactions with the walls and floor of the groove itself, as well as having hydrogen bonds to the N3 atoms of adenines or O2 of thymines. This is direct sequence readout. Indirect sequence readout is via both groove width and complementation of the negative electrostatic potential in the AT minor groove with the positive charge on these drugs. Groove interaction involves the concave curvature of the inner surface of the drug molecule complementing that of the convex surface of the floor of the DNA minor groove itself. This surface matching has been termed isohelicity (29) since both the groove floor and the drug inner surface have twists in their curvature, the former as a result of the helical nature of the DNA double helix. Isohelicity has been found to be a useful concept in the design of novel groove-binding agents.

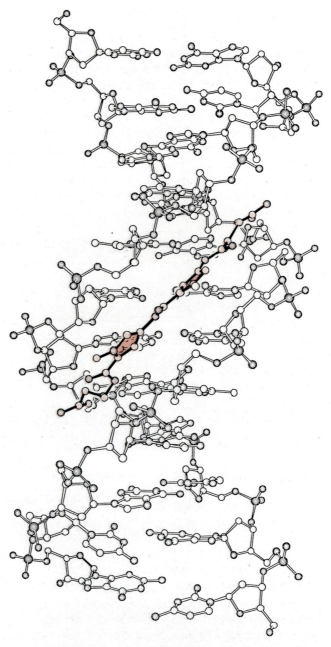

Figure 5.8.   A view of the structure of the netropsin–d(CGCGAATTCGCG)$_2$ complex, from the single-crystal analysis (ref. 27). The drug molecule is shown with shaded bonds.

3.2 Covalent bonding

Non-specific binding to the phosphodiester backbone or sugar residues is presumed to occur with the majority of drugs in this class, although it has only been thoroughly studied when of functional interest, for example when it is a step in DNA strand scission, as in the case of the bleomycin class of drug. There has been most emphasis on recognition of particular sites on the DNA bases by nucleophilic reactions. Purines are the most susceptible to covalent attack, with guanine being preferred over adenine. Particular sites are O6, N6, and N7 in the major groove and N1 and N2 in the minor one. Drugs can bind to a single site or to two at once if it has bifunctional capability. This latter cross-linking mode can be either intra- or inter-strand, depending on two principal factors: (a) the distance between the two functional groups on the drug, and (b) the affected DNA sequence—for example whether two adjacent guanines are on the same or opposite strands. Information on the structures of covalent adducts has to date been obtained on the one hand from chemical and biophysical probe experiments, and on the other, from NMR studies. X-ray crystallography has not as yet had an impact in this area.

Sequence-specific DNA recognition is well illustrated in the case of the anti-tumour antibiotic (+)-CC-1065 (*Figure 5.5*), which has three pyrroloindole units (30). One end of the molecule binds covalently to N3 of an adenine via opening of the cyclopropane ring. The rest of the drug lies in the minor groove, covering four base pairs to the 5′ side of this adenine and one base pair on the 3′ side. Specific recognition is to sequences such as 5′-AAAAA; the drug induces bending of 17–22°, comparable to that found in natural A tracts (Chapter 3), which enables a close steric fit between drug and DNA to take place. So the specific (and highly stringent) sequence requirement is for one that will bend in an appropriate manner.

4. Principles of protein–DNA recognition

4.1 Introduction

DNA-binding proteins can be conveniently categorized in functional terms:

(1) regulatory proteins, which mostly bind to highly specific sequences of duplex DNA in order to control the processing of a particular gene, or bind to particular signal sequences such as 5′-TATA in order to more generally initiate transcription;

(2) DNA cleavage proteins (nucleases)—some, such as DNase I, have relatively little sequence specificity (although they may have some DNA structural selectivity), while others, the restriction enzymes, are highly specific for particular sequences;

(3) repair proteins that respond to various types of damage to DNA by recognizing the lesion itself, then excising the damaged DNA and/or joining together breaks in damaged DNA;

(4) proteins that resolve topological problems in DNA by unravelling or unwinding DNA prior to replication;

(5) structural proteins that maintain the integrity of folded or packaged DNA, for example histones in chromatin;

(6) processing proteins, typified by the polymerases, that use DNA as a template for further nucleic acid synthesis: sequence specificity is absolutely not required for DNA recognition, rather a need to recognize a particular type of DNA duplex structure.

This diversity of functions shown by DNA-binding proteins is in striking contrast to the relatively few ways in which DNA structure is 'read', even though there is very wide variety in the structure of the proteins themselves. One would expect direct reading of a DNA sequence to occur generally via the hydrogen bonding edges of the bases, as described in Section 1 above; therefore features are required of a protein that enable it to gain access to the bases through either major or minor grooves. As has been discussed previously, the major groove is the richer of the two both in information *per se*, and in discrimination between sequences. Thus, it is generally the site of direct information readout. None the less, the minor groove is an important target for some regulatory proteins such as the TFIID TATA-box binding protein, which directly reads the base hydrogen bonding potential in this groove. Indirect readout can occur via the sugar–phosphate backbone or solely the phosphate groups. Especially when it is sequence-specific, it is the cumulative consequence of a number of individually non-specific non-bonded interactions as compared with the clearly defined hydrogen bonding interactions involved in direct recognition. *Table 5.1* details the major recognition motifs found to date in protein–DNA complexes, almost entirely from crystal structures. Undoubtedly more types of motif remain to be discovered, especially among eukaryotic regulatory and damage-repair proteins.

4.2 Major groove interactions—the α-helix as the recognition element

4.2.1 The helix-turn-helix motif

A protein α-helix can fit snugly only into the major groove rather than the minor groove of a B-like DNA duplex and therefore can act directly as the recognition element. This element was first observed in the bacterial *cro* repressor protein (44) as part of the helix-turn-helix (HTH) motif. This HTH domain pattern has subsequently been found in the crystal and solution NMR structures of a number of prokaryotic and eukaryotic transcription regulatory and related proteins, and by comparative sequence analysis in many others. The HTH motif consists of 20 amino acid residues with residues 1–7 forming the first α-helix and residues 12–20 the second, linked by a short turn so that the two helices are inclined at 120° to each other. The second helix is the recognition one, which tends to make most of the specific contacts to DNA and lies in the major groove. There are characteristic hydrophobic residues at positions 4, 8, 10, 16, and 18 of the recognition helix, which help to form the overall hydrophobic core of the protein.

Table 5.1   Recognition motifs found in the crystal structures of some DNA-binding proteins

Protein	Function	DNA sequence recognized	Protein motif	Reference
λ	phage repressor	TATCACCGC	HTH[a]	31
434	phage repressor	ACAAGAAA	HTH[a]	32
trp	*E. coli* repressor	GTACTAGTTA	HTH[a]	33
Met J	*E. coli* repressor	AGACGTCT	β-sheet[a]	34
Engrailed	*Drosophila* gene regulator	TAAT	HTH	35
MATα2	yeast repressor	CATGTAATT	HTH	36
CAP	*E. coli* gene activator	AAAAGTGTGACAT	HTH[a]	37
GAL4	yeast transcription activator	CCGGAGGACAG	zinc-containing domain[a]	38
E2	papillomavirus transcription regulator	ACCGACGTCGGT	β-barrel	39
GCN4	yeast transcription activator	ATGACT	leucine zipper[a]	40
Zif268	murine gene regulator	GCGTGGGCG	zinc finger	
Glucocorticoid receptor	transcription modulator	CAGAACATC	modified zinc[a] finger	42
TFIID	transcription initiation factor	TATATAAA	β-sheet saddle	43

[a] Instances where recognition is by means of a protein dimer. In these cases, only the symmetric or pseudo-symmetric monomer–DNA recognition sequence is given.

The HTH motif does not exist by itself, but is held together in a variety of ways. Thus, the eukaryotic homeodomain (35, 36) has a three-helix bundle (*Figure 5.9a*), with the HTH motif itself occurring after the first helix, whereas the phage λ (31) and 434 repressor (32) structures are five-helix bundles with the HTH motif comprising helices 2 and 3 (*Figure 5.9b*). An important difference between the bacterial and eukaryotic HTH proteins is that the former generally bind to their target sites as dimers whereas the latter bind as monomers. This difference is reflected in the structures of the proteins themselves, with some aspects of the non-HTH domains being responsible for dimerization. These structures as seen in the crystalline state are largely preserved in solution, although side chain orientations are, perhaps unsurprisingly, not always identical (45, 46).

4.2.2 DNA–protein hydrogen bonding and readout: conserved bases and residues

The majority of HTH proteins have a number of direct interactions between the recognition helix and the major groove of the operator sequence DNA, in a B-type conformation (*Figure 5.10*). These interactions involve hydrogen bonds between amino acid side-chains and the edges of the base pairs, involving their

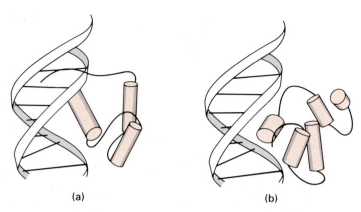

(a)   (b)

Figure 5.9. Schematic views of (a) a homeodomain–DNA complex (ref. 35), and (b) the repressor 434–DNA complex (ref. 32). The α-helices of the proteins are shown as cylinders, with the recognition helix shaded in each case.

pattern of donors and acceptors, as discussed in Section 1 of this chapter (*Figure 5.1*). The protein backbone itself does sometimes make contact with bases or phosphates, but these are generally not determinants of specificity. The majority of interactions involve O6 and/or N7 atoms of guanine bases forming hydrogen bonds with the charged ends of long flexible side-chains from the basic residues arginine or lysine, the amide residues glutamine and asparagine, or the hydroxyl group of a serine. The recognition of one hydrogen bonding site on a base, compared with two simultaneous sites on that base, actually involves only a small change in position of the amino acid side-chain. In general, the latter will be energetically preferred and formed if possible. Adenine bases are recognized via their N6 and/or N7 atoms, although this occurs much less frequently than guanine recognition; active pyrimidine recognition is even less common. However, there is no simple 1:1 amino acid:DNA base correspondence, since recognition can occur in a wide variety of ways in addition to the simple mono- or bidentate ones (*Figure 5.10*). For example, it is common to find side-chains bridging between two adjacent bases, which can be on the same strand or even on opposite strands, with each base being bidentate hydrogen bonded to the amino acid. Sometimes a water molecule (or molecules) participates in such a hydrogen bonded bridge, or itself bridges between a base and a side-chain. Such water involvement has been found in the *E. coli trp* repressor–operator complex (33), with only two (relatively unimportant) direct (guanine. . .arginine) contacts, yet three that have water-mediated contacts. A further example of the important role that water molecules can play is in the bacteriophage λ repressor–operator complex (31) where the hydroxyl group of serine-45 interacts directly with O6

Figure 5.10. (a) Schematic representation of several modes of direct base readout from α-helical side chains of arginine, glutamine, and lysine residues to adenine and guanine bases. (b) The structure of the MATα2–DNA complex (ref. 36), viewed down the recognition α-helix and showing its amino-acid side-chains.

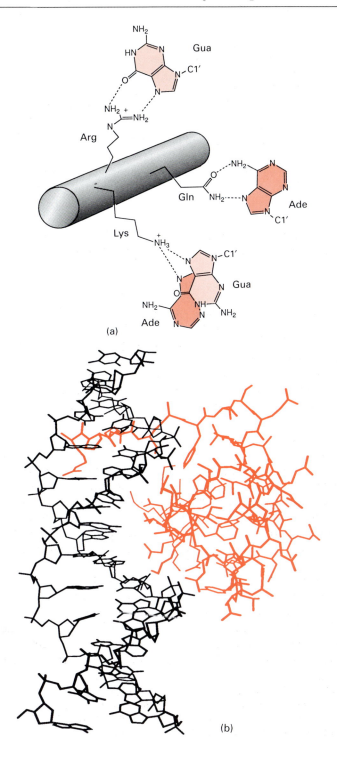

(a)

(b)

and N7 of a guanine, and the backbone carbonyl oxygen atom of this serine interacts via a water molecule with the N4 atom of the complementary cytosine base and through the same water molecule to O4 of the adjacent thymine.

It has not been possible to establish a pattern of preference between a particular base and an amino acid residue. This lack of a general hydrogen bonding recognition pattern (although there are some involving very closely related proteins), reflects in part the effect of differences in DNA structure in various complexes. Other recognition factors can play significant roles in defining the recognition of particular sequences, notably the ability of thymine methyl groups to form close van der Waals attractive interactions with hydrophobic side chains. An example of this is in the structure of the *Drosophila* engrailed homeodomain, where there is a close contact between isoleucine-47 and a methyl group from a thymine in the TAAT recognition site (35). At least as important roles are played by numerous non-specific (at least in hydrogen bonding terms) protein–backbone and other interactions in stabilizing a protein–DNA complex as a whole. Thus the flexibility of amino acid side-chains ensures that optimal interactions are made with all elements of a DNA structure. Taken as a whole, these enable the recognition helix to be oriented in the major groove in a manner that is distinctive for a particular protein. Such generalized hydrophobic and electrostatic interactions may also play a more active recognition role, by means of indirect readout, sensing the DNA structure and flexibility specified by a particular sequence. In view of all of these factors it is then unsurprising that there is wide variation in the number and importance of direct readout interactions that have been observed in different HTH complexes. The major role that can be played by indirect readout is strikingly demonstrated in the structure of the *trp* repressor–operator complex (33), which has a large number of contacts between side-chains and the phosphate groups of the operator DNA, yet very surprisingly has no functionally significant direct readout base–amino acid interactions at all. Instead the cumulative effect of the indirect readout contacts made by the *trp* repressor is to read effectively the particular structural details of its DNA operator sequence (47).

Readout, both direct and indirect, ensures that key residues on both protein and DNA are effectively recognized. The functional importance of such residues, and hence the protein–DNA contacts involved, can be evaluated by mutagenesis experiments, as well as by surveying their occurrence within a particular family of proteins and their consensus operator sequences. For example, residues tryptophan-48, phenylalanine-49, asparagine-51, and arginine-53 occur in every member of the large eukaryotic homeodomain family; the crystal structure of the engrailed homeodomain–DNA complex (35) shows that the first two play critical roles in preserving the hydrophobic core of the recognition helix, and that the other two directly interact with base and phosphate respectively.

4.3 Zinc finger and other recognition modes

Several other motifs for sequence-specific recognition have been envisaged, and their details revealed by crystallographic and NMR studies. The largest group in

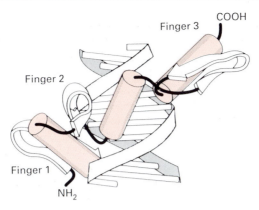

Figure 5.11. Schematic representation of the structure of the zinc-finger protein Zif268 complexed to a DNA sequence. Adapted from ref 41.

the family of zinc-containing DNA-binding proteins, are the zinc-finger proteins, which contain units of regularly spaced cysteine and histidine residues co-ordinated to a zinc ion. Each finger consists of an anti-parallel β-sheet and an α-helix, in part held together by the zinc ion coordination (41). Zinc-finger proteins have at least two such units ('fingers'), which each recognize a 3 bp site in the major groove of a B-DNA helix (*Figure 5.11*). The finger makes direct contacts to the guanine-rich strand, with patterns of side-chain interactions from the α-helix involving arginine . . . guanine recognition very similar to those seen in HTH proteins. A quite distinct zinc domain is involved in the structure of the yeast transcription activator GAL4 (38), which binds as a dimer. Each individual GAL4 molecule has two domains, a zinc-binding region and an α-helix, linked by an extended length of peptide (*Figure 5.12*). The zinc-containing domain rests in the DNA major groove, where there is a pattern of direct interactions between basic lysine side-chains and guanine/cytosine bases that exactly specifies the highly conserved sequence CCG at this point along the DNA.

A number of eukaryotic transcription factors contain the basic 'leucine-zipper' recognition and dimerization element (40), which consists of a very long, continuous, almost straight α-helix with regularly repeating leucine residues. Two such helices interact together to form a parallel coiled coil. There are a large number of interactions with the DNA backbone from the basic side-chains, as well as specific, direct readout ones. The orientation of the α-helix within the DNA major groove is critical for these contacts to take place, with the helix being approximately parallel to the phosphodiester backbone.

A hitherto unpredicted DNA-binding motif is used by the transcription activator E2 from papilloma virus (39), with a dimeric anti-parallel β-barrel which delivers a pair of α-helices to the major groove. Protein–DNA contacts involve direct side-chain interactions with the bases, as well as indirect backbone contacts. A distinguishing feature of the structure is the smooth bending of the DNA around the β-barrel, with a radius of curvature of 45 Å. This is less extreme than the

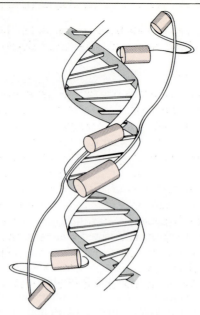

Figure 5.12.   A representation of the GAL4–DNA molecular structure. Adapted from ref 38.

bend found (39) in the DNA of the complex with the dimeric HTH domain *E. coli* transcription activator protein (CAP). The bending is due to a large number of interactions with DNA phosphate groups, which serve to position the recognition helix correctly in the major groove (*Figure 5.13*). A totally distinct non-helix recognition motif has been found in the structures of a few bacterial repressors, typified by the *met* J repressor–operator complex (34), where double-stranded anti-parallel β ribbons are the major groove recognition elements; side-chains from the ribbons interact directly with the DNA phosphates and bases, in ways analogous to those observed with the other repressor–operator complexes

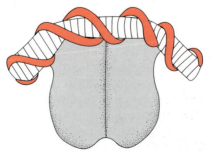

Figure 5.13.   Representation of the CAP–DNA complex, showing the protein dimer and the DNA bent around it.

outlined above. Other types of folds have been found in the structures of the few DNA repair enzymes whose structures have been determined.

4.4 Minor groove recognition

Sequence-specific proteins generally exploit the major groove to a greater extent than the minor one. This is not always solely on account of its greater information potential. The ability of an α-helix to fit snugly into the major groove is also a factor. This led in the past to a view that the minor groove is of little importance. However, there are now a number of examples where interaction of side-chains in the minor groove is a significant component of the overall protein–DNA stabilization. An arginine-side chain of the 434 repressor (32) bridges to bases and phosphate groups via water molecules. The two established homeodomain structures (34, 35) have extended N-terminal arms which lie in the DNA minor groove (*Figures 5.10b, 5.14*), making extensive base and backbone contacts to the AT regions of the operator sequence. The hydrogen bonding interactions that the arginine side-chains in these complexes make with the O2 atom of thymines is closely analogous to the mode of interaction shown by the minor-

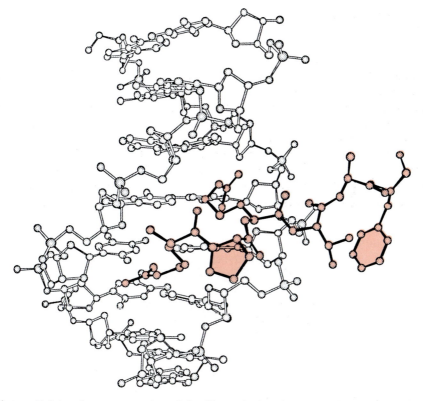

Figure 5.14.   A computer plot of the N-terminal region of the homeodomain–DNA molecular structure (ref. 35), with the protein backbone and side-chains shown in bold.

groove binding drugs (see Section 3.1 above). These, together with the indirect readout of the dimensions of the groove itself by van der Waals interactions involving the side-chains, are a significant factor in the preference shown by homeodomains for AT-rich sites on DNA, again analogous to the preferences shown by the drugs. These N-terminal sequences are also related to the proposal (48) that the repeating sequence SPKK (serine-proline-lysine/arginine-lysine/arginine), which occurs in, for example, the N-terminus of some histone proteins, also binds in the minor groove at AT-rich regions.

The TFIID protein is a key factor in the initiation of transcription in eukaryotic cells. It functions by binding to the 'TATA box' sequence upstream of the start of transcription, and has been shown by chemical protection studies to bind in the minor groove at the TATA site itself, as well as in the major groove on the 3′ side of this site. There is a strong preference for TATA as compared with other AT-containing sequences, by contrast with the weak preferences shown by typical minor-groove drugs. The crystal structure of TFIID both alone (43a) and bound to DNA (43b), shows a novel DNA-binding fold, with a symmetric α/β structure (*Figure 5.15*). This saddle-shaped arrangement has an extended concave surface with this minor groove of the DNA being shaped around this surface

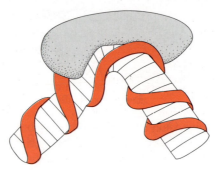

Figure 5.15.   A representation of the TFIID-DNA structure, showing the position of the bound DNA duplex. Adapted from ref. 43b.

and consequently severely bent towards the major groove. The DNA is bent by about 80° as a consequence. Mutagenesis studies have identified the key protein residues involved in DNA binding. They are arranged on the concave surface of the protein, in agreement with the structures of the DNA complexes. The ability of TFIID to bend DNA on binding to its recognition sequence is shared by a number of other minor-groove regulatory proteins, notably those containing the so-called HMG (high mobility group) box domains. Such proteins include SRY, which determines the expression of human male-specific genes. An NMR structure of an HMG box (49) shows that the structure consists of three α-helices arranged in an L shape, with conserved basic arginines and lysines on the inner face of the L. The helices do not correspond to those in any (major-groove binding) HTH arrangement, and thus constitute a novel DNA-binding motif.

5. DNA conformation and drug/protein recognition

X-ray crystallography and other techniques have shown that DNA structures in the absence of bound drug or protein have a significant degree of flexibility that is reflected in sequence-dependent backbone, sugar pucker, base step, and base pair features. The details of the classic exactly repetitious fibre diffraction forms are rarely if ever observed in native oligonucleotide structures.

Are these features observed in drug and protein complexes? The (non-oligonucleotide) groove-binding drugs do not need to distort DNA appreciably from a B-form, and thus the crystal and solution structures of their complexes show oligonucleotides essentially as in their drug-free states. As yet there are no sufficiently high-resolution pictures of triple-stranded DNAs to indicate clearly their detailed structures; however, it seems likely that aspects of both A- and B-type structures will be involved.

The situation with proteins is quite different, largely because the binding energy for a protein–DNA interaction is sufficient to overcome the barriers to local DNA deformability, and hence can lead to even large-scale changes in DNA structure such as bending, so as to optimize interactions. Sometimes these distortions blur the distinctions between B- and A-form DNA structural types so that it is no longer possible unequivocally to assign a given DNA operator structure as one or the other, especially when changes in parameters such as groove width, helical twist, and base pair inclination to the helix axis are considered. For example, the nature of the structure of DNA in a zinc-finger complex has been controversial, with both A- and B-type helices being suggested. The structure of the Zif268–DNA complex (41) shows that the DNA is overall of B-type, although there are a number of local features that are intermediate between A and B. In such a case, the simple canonical DNA classifications are of limited use by themselves and are best accompanied by more detailed information describing particular features.

The DNA in the *met* J complex (34) has a straight 10 bp central segment, with bends at each end as a result of groove width changes necessary for both effective protein dimer formation and protein–DNA interaction. In the structure of the *trp* repressor–operator complex (33) the DNA is appreciably distorted from canonical B-form as a result of numerous small, cumulative local changes to, in particular, roll and slide. These together with changes in groove width and backbone geometry from canonical B-DNA are necessary in order to achieve the large number of indirect readout contacts that produce specificity. The DNA is bent so as to follow closely the contours of the protein surface. Rather greater DNA bending is required in the papillomavirus-1 E2–DNA complex (39), where there is smooth DNA bending over the β-barrel recognition motif, in order to expose the major groove to protein side-chains. Groove widths are compressed on the side of the DNA that faces protein, with alterations in propeller twist and roll angles contributing to these width changes.

The prokaryotic transcription activator CAP, which acts as a dimer, has a consensus DNA-binding site, 22 bp long, with the essential bases towards the end of the sequence. This DNA sequence would be far too long to be contacted by the relatively small CAP dimer if the DNA remained linear. The structure of the CAP–DNA complex shows (37) that this problem has been solved by the DNA bending by ~90° in order to wrap effectively around the protein dimer. This bending, or kinking, is produced almost entirely at two bp steps, one on each side of the 2-fold axis of the complex, by roll angles of ~40°. It is likely that this bending is an important aspect of transcriptional activation by CAP and other analogous proteins. There is a very close structural correspondence between CAP and the globular domain of histone H5 (50) (which is involved in nucleosome organization), suggesting that the ability of this particular type of HTH protein to bend DNA sequences is of even wider functional significance.

6. Further reading

DNA–water interactions:

Berman,H.M. (1991). *Current Opinion in Structural Biology*, **1**, 423.
Westhof,E. (1988). *Annual Review of Biophysics*, **17**, 125.
Westhof,E. and Beveridge,D.L. (1990). In *Water science reviews 5* (ed. F. Franks), pp. 24–136. Cambridge University Press, Cambridge, UK.

DNA–drug interactions:

Gale,E.F., Cundliffe,E.F., Reynolds,P.E., Richmond,M.H., and Waring,M.J. (1981). *The molecular basis of antibiotic action.* John Wiley, London.
Gao,X. and Patel,D.J. (1989). *Quarterly Review of Biophysics*, **22**, 93.
Hurley,L.H. (1989). *Journal of Medicinal Chemistry*, **30**, 2027.
Kopka,M.L. and Larsen,T.A. (1992). In *Nucleic acid targeted drug design* (ed. C.L.Propst, and T.J.Perun), pp. 304–74. Marcel Dekker, New York.
Neidle,S. and Waring,M.J. (eds) (1993). *Molecular basis of drug–DNA antitumour action*, Vols 1 and 2. Macmillan Press, London.
Pullman,B. (1989). *Advances in Drug Research*, **18**, 1.

DNA–protein interactions:

Freemont,P.S., Lane,A.N., and Sanderson,M.R. (1991). *Biochemical Journal*, **278**, 1.
Harrison,S.C. (1991). *Nature*, **353**, 715.
Harrison,S.C. and Aggarwal, A.K. (1990). *Annual Review of Biochemistry*, **59**, 933.
Pabo,C.O. and Sauer,R.T. (1992). *Annual Review of Biochemistry*, **61**, 1053.
Steitz,T.A. (1990). *Quarterly Review of Biophysics*, **23**, 205.
Travers,A.A. (1993). *DNA-protein interactions.* Chapman & Hall, London.

7. References

1. Seeman,N.C., Rosenberg,J.M., and Rich,A. (1976). *Proceedings of the National Academy of Sciences, USA*, **73**, 804.
2. Pullman,A. and Pullman,B. (1981). *Quarterly Review of Biophysics*, **14**, 289
3. Kuninec,M.G. and Wemmer,D.E. (1992). *Journal of the American Chemical Society*, **114**, 8739.
4. Liepinsh,E., Otting,G., and Wüthrich,K. (1992). *Nucleic Acids Research*, **20**, 6549.

5. Alden, C.J. and Kim, S.-H. (1979). *Journal of Molecular Biology*, **132**, 411.
6. Schneider, B., Cohen, D., and Berman, H.M. (1992). *Biopolymers*, **32**, 725.
7. Saenger, W., Hunter, W.N., and Kennard, O. (1986). *Nature*, **324**, 385.
8. Kennard, O., Cruse, W.B.T., Nachman, J., Prange, T., Shakked, Z., and Rabinovich, D. (1986). *Journal of Biomolecular Structure and Dynamics*, **3**, 623.
9. Lerman, L.S. (1961). *Journal of Molecular Biology*, **3**, 18.
10. Waring, M.J. (1970). *Journal of Molecular Biology*, **54**, 247.
11. Neidle, S., Achari, A., Taylor, G.L., Berman, H.M., Carrell, H.L., Glusker, J.P., and Stallings, W.C. (1977). *Nature*, **269**, 304.
12. Wang, A.H.-J., Ughetto, G., Quigley, G.J., and Rich, A. (1987). *Biochemistry*, **26**, 1152.
13. Kamitori, S. and Takusagawa, F. (1992). *Journal of Molecular Biology*, **225**, 445.
14. Liu, X., Chen, H., and Patel, D.J. (1991). *Journal of Biomolecular NMR*, **1**, 323.
15. Egli, M., Williams, L.D., Frederick, C.A., and Rich, A. (1991). *Biochemistry*, **30**, 1364.
16. Zhang, X. and Patel, D.J. (1990). *Biochemistry*, **29**, 9451.
17. Fox, K.R. and Waring, M.J. (1986). *Biochemistry*, **25**, 4349.
18. Wang, A.H.-J., Ughetto, G., Quigley, G.J., Hakoshima, T., van der Marel, G.A., van Boom, J.H., and Rich, A. (1984). *Science*, **225**, 1115.
19. Quigley, G.J., Ughetto, G., van der Marel, G.A., van Boom, J.H., Wang, A.H.-J., and Rich, A. (1986). *Science*, **232**, 1255.
20. Gao, X. and Patel, D.J. (1988). *Biochemistry*, **27**, 1744.
21. Gilbert, D.E. and Feigon, J. (1992). *Nucleic Acids Research*, **20**, 2411.
22. McLean, M.J., Seela, F., and Waring, M.J. (1989). *Proceedings of the National Academy of Sciences, USA*, **86**, 9687.
23. Zimmer, C.H. and Wähnert, U. (1986). *Progress in Biophysics and Molecular Biology*, **47**, 31.
24. Dervan, P.B. (1986). *Science*, **232**, 464.
25. Brown, D.G., Sanderson, M.R., Garman, E., and Neidle, S. (1992). *Journal of Molecular Biology*, **226**, 481.
26. Lane, A.N., Jenkins, T.C., Brown, T., and Neidle, S. (1991). *Biochemistry*, **30**, 1372.
27. Kopka, M.L., Yoon, C., Goodsell, D., Pjura, P., and Dickerson, R.E. (1986). *Journal of Molecular Biology*, **183**, 553.
28. Patel, D.J. and Shapiro, L. (1986). *Journal of Biological Chemistry*, **261**, 1230.
29. Goodsell, D. and Dickerson, R.E. (1986). *Journal of Medicinal Chemistry*, **29**, 727.
30. Hurley, L.H., Lee, C.-S., McGovren, J.P., Mitchell, M.A., Warpehoski, M.A., Kelly, R.C., and Aristoff, P.A. (1988). *Biochemistry*, **27**, 3886.
31. Beamer, L.J. and Pabo, C.O. (1992). *Journal of Molecular Biology*, **227**, 177.
32. Aggarwal, A.K., Rodgers, D.W., Drottar, M., Ptashne, M., and Harrison, S.C. (1988). *Science*, **242**, 899.
33. Otwinowski, Z., Schevitz, R.W., Zhang, R.-G., Lawson, C.L., Joachimiak, A., Marmorstein, R.Q., Luisi, B.F., and Sigler, P.B. (1988). *Nature*, **335**, 321.
34. Somers, W.S. and Phillips, S.E.V. (1992). *Nature*, **359**, 387.
35. Kissinger, C.R., Liu, B., Martin-Blanco, E., Kornberg, T.B., and Pabo, C.O. (1990). *Cell*, **63**, 579.
36. Wolberger, C., Vershon, A.K., Liu, B., Johnson, A.D., and Pabo, C.O. (1991). *Cell*, **67**, 517.
37. Schultz, S.C., Shields, G.C., and Steitz, T.A. (1991). *Science*, **253**, 1001.
38. Marmorstein, R., Carey, M., Ptashne, M., and Harrison, S.C. (1992). *Nature*, **356**, 408.
39. Hegde, R.S., Grossman, S.R., Laimonis, L.A., and Sigler, P.B. (1992). *Nature*, **359**, 505.
40. Ellenberger, T.E., Brandl, C.J., Struhl, K., and Harrison, S.C. (1992). *Cell*, **71**, 1223.

41. Pavletich,N.P. and Pabo,C.O. (1991). *Science,* **252,** 809.
42. Luisi,B., Xu,W.X., Freedman,L.P., Yamamato,K.R., and Sigler,P.B. (1991). *Nature,* **352,** 497.
43. (a) Nikolov,D.B., Hu,S.-H., Lin,J., Gasch,A., Hoffmann,A., Horikoshi,M., Chua, N.-H., Roeder,R.G., and Burley,S.K. (1992). *Nature,* **360,** 40 (b) Kim,Y., Geiger,J.H., Hahn,S., and Sigler,P.B., (1993). *Nature,* **365,** 512; Kim,J.L., Nikolov,D.B., and Burley,S.K. (1993). *Nature,* **365,** 520.
44. Brennan,R.G. (1991). *Current Opinion in Structural Biology,* **1,** 80.
45. Neri,D., Billeter,M., and Wüthrich,K. (1992). *Journal of Molecular Biology,* **223,** 743.
46. Kaptein,R. (1993). *Current Opinion in Structural Biology,* **3,** 50.
47. Haran,T.E., Joachimiak,A., and Sigler,P.B. (1992). *The EMBO Journal,* **11,** 3021.
48. Suzuki,M. (1988). The *EMBO Journal,* **8,** 797.
49. Weir,H.M., Kraulis,P.J., Hill,C.S., Raine,A.R.C., Laue,E.D., and Thomas,J.O. (1993). The *EMBO Journal,* **12,** 1311.
50. Ramakrishnan,V., Finch,J.T., Graziano,V., Lee,P.L., and Sweet,R.M. (1993). *Nature,* **362,** 219.

A1

Detailed aspects of DNA structure

Table A.1 provides numerical data for the major conformational classes of DNA, as polymers in fibres, and as oligonucleotides in single crystals. Data on individual structures are obtainable from the Brookhaven and Nucleic Acid Databases.

A standard convention has been developed that defines the various local base and base pair parameters (3, 4) and a computer program (from the Brookhaven Data Bank: 'NEWHEL', written by R.E. Dickerson) is available that calculates them as well as other global helical parameters, given a set of double-helical nucleic acid or oligonucleotide coordinates. This program calculates a least-squares best-fit to a straight-line helix from the coordinate data. This assumption of a linear helix axis is rarely true in practice. Other more sophisticated procedures have been developed that do take helix curvature into account (5, 6). The CURVES program (5) finds use in the analysis of structures with pronounced DNA bending.

Table A.1 Conformational angles for several of the major polymorphic structures of DNA

Polymorph		α	β	γ	δ	ε	ζ	χ
(a) Values from fibre-diffraction analyses (taken from ref. 1)								
A		−52	175	42	79	−148	−75	−157
B		−30	136	31	143	−141	−161	−98
C		−37	−160	37	157	161	−106	−97
D	A	−51	140	61	146	−128	−141	−115
	T	−76	142	68	148	−152	−154	−105
Z	G	52	179	−174	95	−104	−65	59
	C	−140	−137	51	138	−97	82	−154
(b) Averaged values from single-crystal studies (from ref. 2)								
A		−73	173	64	78	−151	−77	−165
B		−65	167	51	129	−157	−120	−103
Z	G	48	179	−170	100	−104	−69	67
	C	−137	−139	55	138	−94	80	−159

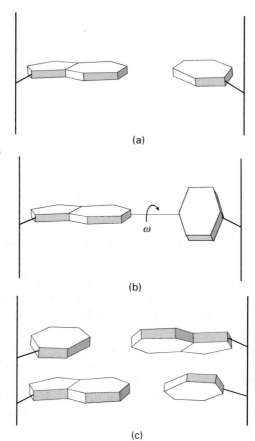

(a)

(b)

(c)

Figure A.1.   Schematic views of a base pair with (a) zero and (b) high propeller twist. Panel (c) shows the effects of propeller twist on two successive base pairs.

The principal rotational and translational parameters that describe relations between bases and base pairs are as follows.

(1) For individual base pairs:

    (a) **Propeller twist** (ω) between bases is the dihedral angle between normals to the bases, when viewed along the long axis of the base pair (*Figure A.1*). The angle has a negative sign under normal circumstances, with a clockwise rotation of the nearer base when viewed down the long axis. The long axis for a purine–pyrimidine base pair is defined as the vector between the C8 atom of the purine and the C6 of a pyrimidine in a Watson–Crick base pair. Analogous definitions can be applied to other non-standard base pairings in a duplex including purine–purine and pyrimidine–pyrimidine ones.

    (b) **Buckle** (κ) is the dihedral angle between bases, along their short axis, after propeller twist has been set to 0° (*Figure A.2*). The sign of buckle

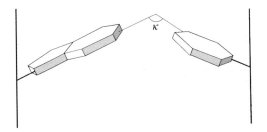

Figure A.2.   View of base-pair buckle.

is defined as positive if the distortion is convex in the direction 5′→3′ of strand 1. The change in buckle for succeeding steps, termed **cup**, has been found to be a useful measure of changes along a sequence. Cup is defined as the difference between the buckle at a given step, and that of the preceding one.

(c) **Inclination** (η) is the angle between the long axis of a base pair and a plane perpendicular to the helix axis. This angle is defined as positive for right-handed rotation about a vector from the helix axis towards the major groove.

(d) **X** and **Y** **displacements** define translations of a base pair within its mean plane in terms of the distance of the mid-point of the base pair long axis from the helix axis. X displacement is towards the major groove direction, when it has a positive value. Y displacement is orthogonal to this, and is positive if towards the first nucleic acid strand of the duplex.

(2) For base pair steps:

(e) **Helical twist** (Ω) is the angle between successive base pairs, measured as the change in orientation of the C1′–C1′ vectors on going from one base pair to the next, projected down the helix axis (*Figure A.3*). For an exactly repetitious double helix, helical twist is 360°/n, where n is the unit repeat defined above.

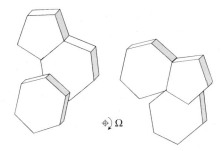

Figure A.3.   View down two successive base pairs, showing the helical twist angle between them.

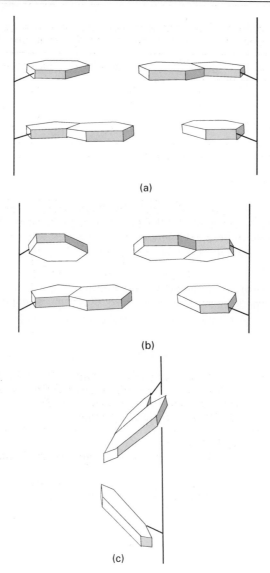

Figure A.4.   Views of two successive base pairs (a) with 0° roll angle between them, and (b) with a positive roll angle. Panel (c) shows a view of positive roll along the long axis of the base pairs.

(f) **Roll** (ρ) is the dihedral angle for rotation of one base pair with respect its neighbour, about the long axis of the base pair. A positive roll angle opens up a base pair step towards the minor groove (*Figure A.4*).

(g) **Slide** is the relative displacement of one base pair compared with another, in the direction of nucleic acid strand 1 (i.e. the Y displacement), measured between the mid-points of each C6–C8 base pair long axis.

Table A.2   (a) Selected base pair and base step morphological parameters in fibrous polynucleotide structures (from ref. 1), as defined in Section A.1 above

Helix	Propeller twist (°)	Roll (°)	Helical twist (°)	Inclination (°)	Rise (Å)
A	−10	0.0	32	23	2.5
B	−15	0.0	36	8	3.4
C	−2	0.0	39	8	3.3
D*	−17, −17	2, −2	45, 45	18, 16	6.1

The pairs of values for the Z helix are for G and C residues respectively in this dinucleotide repeating polynucleotide.
* denotes a dinucleotide repeat

(b) Averaged from four B-DNA decamer, six A-DNA octamer single-crystal oligo-nucleotide structures and the Z_I-DNA hexamer structure (adapted from ref. 2)

A	−8	6	31	12	2.9
B	−11	1	36	2	3.4
Z: C	−1	−6 (CpG)	−9 (CpG)	−6	3.9 (CpG)
G	−1	6 (GpC)	−51 (GpC)	−6	3.5 (GpC)

References

1. Chandrasekaran,R. and Arnott,S. (1989). In *Landolt-Börnstein, New series*, Group VII, Vol. 1b (ed. W. Saenger), pp. 68–85. Springer-Verlag, Berlin.
2. Dickerson,R.E. (1992). *Methods in Enzymology*, **211**, 67.
3. Dickerson,R.E. *et al.* (1989). *The EMBO Journal*, **8**, 1.
4. Fratini,A.V., Kopka,M.L., Drew,H.R., and Dickerson,R.E. (1982). *Journal of Biological Chemistry*, **257**, 14686.
5. Lavery,R. and Sklenar,H. (1989). *Journal of Biomolecular Structure and Dynamics*, **6**, 655.
6. Badcock,M.S. and Olson,W.K. (1993). In *Computation of biomolecular structure* (ed. D.M.Soumpasis and T.M.Jovin). Springer-Verlag, Berlin.

Glossary

A-DNA: A right-handed anti-parallel duplex with the bases inclined to the helix axis and significantly displaced from it.

A tract: a sequence of DNA double helix with solely adenines on one strand.

B-DNA: an anti-parallel right-handed duplex with the bases approximately perpendicular to the helix axis and having only a small displacement from it.

Base pairing: hydrogen bonding interactions between purine and/or pyrimidine bases.

Direct readout: recognition of a particular base pair or a DNA sequence by means of hydrogen bonding to donors and/or acceptors on a base pair edge.

Glycosidic angle: the torsion angle between base and deoxyribose sugar, around the glycosidic bond linking the two groups.

G-quartet: an arrangement of four guanine bases hydrogen bonded together, all in the same plane.

Grooves: the indentations in the surface of a double helix that arise from the progression of bases and backbone around the helix axis. Their geometry and nature depend on the type of helix.

Indirect readout: recognition of a particular base pair or a DNA sequence by non-bonded interactions with the phosphodiester backbone or other indirect means.

Hoogsteen base pairing: base pairing arrangements between adenine and thymine, and guanine and cytosine, that involve the N7 atom of the purine in hydrogen bonding.

Intercalation: binding of a planar aromatic chromophore to double-stranded DNA by means of stacking in between adjacent base pairs so as to lengthen and unwind a double helix.

Mis-pairing: non-Watson-Crick base pairing.

Sugar pucker: the out-of-plane displacement of atoms within a deoxyribose sugar ring that arise as a consequence of non-bonded interactions.

Triplex: a triple-stranded helix involving a Watson–Crick duplex with conventionally all purines on one strand and all pyrimidines in the other. The third strand is normally either all-pyrimidine or all-purine.

Watson–Crick base pairing: specific hydrogen bonded association between adenine and thymine, or between cytosine and guanine bases; the two types are isomorphous and are present in right-handed anti-parallel DNA duplexes.

Z-DNA: a left-handed anti-parallel helix with alternating purine and pyrimidine nucleosides.

Index

QMW LIBRARY
(MILE END)